INTRODUCTION TO SEARCH ENGINE OPTIMIZATION

A GUIDE FOR ABSOLUTE BEGINNERS

Todd Kelsey

Apre

D0813260

Introduction to Search Engine Optimization: A Guide for Absolute Beginners

Todd Kelsey
Wheaton, Illinois, USA

ISBN-13 (pbk): 978-1-4842-2850-0   ISBN-13 (electronic): 978-1-4842-2851-7
DOI 10.1007/978-1-4842-2851-7

Library of Congress Control Number: 2017945753

Managing Director: Welmoed Spahr
Editorial Director: Todd Green
Acquisitions Editor: Susan McDermott
Development Editor: Laura Berendson
Technical Reviewer: Brandon Lyon
Coordinating Editor: Rita Fernando
Copy Editor: Kezia Endsley
Cover: eStudio Calamar

Distributed to the book trade worldwide by Springer Science+Business Media New York, 233 Spring Street, 6th Floor, New York, NY 10013. Phone 1-800-SPRINGER, fax (201) 348-4505, e-mail orders-ny@springer sbm.com, or visit www.springeronline.com. Apress Media, LLC is a California LLC and the sole member (owner) is Springer Science + Business Media Finance Inc (SSBM Finance Inc). SSBM Finance Inc is a **Delaware** corporation.

For information on translations, please e-mail rights@apress.com, or visit http://www.apress.com/rights-permissions.

Apress titles may be purchased in bulk for academic, corporate, or promotional use. eBook versions and licenses are also available for most titles. For more information, reference our Print and eBook Bulk Sales web page at http://www.apress.com/bulk-sales.

Any source code or other supplementary material referenced by the author in this book is available to readers on GitHub via the book's product page, located at www.apress.com/9781484228500. For more detailed information, please visit http://www.apress.com/source-code.

Apress Business: The Unbiased Source of Business Information

Apress business books provide essential information and practical advice, each written for practitioners by recognized experts. Busy managers and professionals in all areas of the business world—and at all levels of technical sophistication—look to our books for the actionable ideas and tools they need to solve problems, update and enhance their professional skills, make their work lives easier, and capitalize on opportunity.

Whatever the topic on the business spectrum—entrepreneurship, finance, sales, marketing, management, regulation, information technology, among others—Apress has been praised for providing the objective information and unbiased advice you need to excel in your daily work life. Our authors have no axes to grind; they understand they have one job only—to deliver up-to-date, accurate information simply, concisely, and with deep insight that addresses the real needs of our readers.

It is increasingly hard to find information—whether in the news media, on the Internet, and now all too often in books—that is even-handed and has your best interests at heart. We therefore hope that you enjoy this book, which has been carefully crafted to meet our standards of quality and unbiased coverage.

We are always interested in your feedback or ideas for new titles. Perhaps you'd even like to write a book yourself. Whatever the case, reach out to us at editorial@apress.com and an editor will respond swiftly. Incidentally, at the back of this book, you will find a list of useful related titles. Please visit us at www.apress.com to sign up for newsletters and discounts on future purchases.

—*The Apress Business Team*

Contents

About the Author

Todd Kelsey, PhD, is an author and educator whose publishing credits include several books for helping people learn more about technology. He has appeared on television as a featured expert, and has worked with a wide variety of corporations and non-profit organizations. He is currently an Assistant Professor of Marketing at Benedictine University in Lisle, IL (www.ben.edu).

Here's a picture of one of the things I like to do when I'm not doing digital marketing—grow sunflowers! (And measure them. Now there's some analytics for you!)

I've worked professionally in digital marketing for some time now, and I've also authored books on related topics. You're welcome to look me up on LinkedIn, and you're also welcome to invite me to connect: http://linkedin.com/in/tekelsey.

About the Technical Reviewer

Brandon Lyon is an expert in SEO, SEM, and Social Media and Web analytics, and President of Eagle Digital Marketing (https://www.eagle-digitalmarketing.com), a full-service agency in the Chicago area. When he isn't advising local business owners and CEOs of mid-sized companies, he enjoys hockey and doing his best to survive the occasional subzero temperatures. Brandon enjoys helping companies face the challenges of the future with optimism, including navigating the treacherous waters of the Amazon e-commerce river, and taking advantage of the goldmine in marketing automation.

Introduction

Welcome to search engine optimization!

The purpose of this book is to provide a simple, focused introduction to search engine optimization, for interns who may be working at a company or non-profit organization, for students at a university, or for self-paced learners. The approach is the same one taken in most of the books I've written, which is conversational, friendly, with an attempt at making things fun.

The experiment is to help you get started with search engine optimization in a way that is fun and helps strengthen your career at your employer, or help you find work through an internship, paid work, volunteer efforts, freelancing, or any other type of work. The focus is on skills and approaches that might be immediately useful to a business or non-profit organization. I'm not going to try to cover everything—just the things that I think are the most helpful.

The other goal is to help you leave intimidation in the dust. I used to be intimidated by marketing and now look at me, I'm a marketing strategist, and an assistant professor of marketing! But I remember the feeling of intimidation, so part of my approach is to encourage readers who may feel uncertain about the field.

The fact is, that search engine optimization has a lot of option—especially in the "tools" area—it has grown rapidly, and there is a lot of material out there. It can be overwhelming! But it can also be very doable, if you leave intimidation in the dust, by taking incremental steps, trying things out, and building confidence.

For example, I had a friend who used to be a journalist and was looking for career opportunities. I helped get him started in digital marketing, and one of the first things he ran into was feeling overwhelmed by all the options, including all the articles about all the options. "There are so many tools out there!" he used to say, "How am I ever going to learn all of them?!?"

The answer is that you don't need to learn all of them. No one can.

The thing to do is focus on some of the tools and skills and build from there.

I encouraged my friend not to worry about trying to learn everything, but instead to just learn some basics.

The friend worked with Facebook advertising, learned a bit about Twitter, and was able to find a local agency that gave him a shot at doing some freelance digital marketing work. The career didn't develop easily for him—he had to put effort into it. But a few years later, he's doing full-time freelance work in digital marketing and is writing Google ads. He was able to leave his intimidation in the dust—and I believe he's also had some fun with it too.

This book is not a collection of individual skills, such as "The Top 10 List of Most Effective SEO Techniques" or anything like that. It's not comprehensive. A blog search will reveal that kind of material, which can have its uses.

What I try to do is start a conversation and show you some skills, concepts, and topics that have jumped out at me based on my experience. In other words, it's fairly informal and is mainly intended for beginners, especially those who might be intimidated by technology. I think my perspective is valid, but I don't claim to know everything, and I learn from third parties and students all the time. This especially applies to search engine optimization, which changes a lot.

LinkedIn shows digital/online marketing as a top skill to have year after year, and search engine marketing is one of the core skills for digital marketing. It's what makes Google and other companies tens of billions of dollars a year.

Each year the way they refer to digital marketing seems to change, but since 2013, digital marketing (of which analytics is a core part) has been at the top. Demand will fluctuate over time, but these are the top skills in any field that get people hired.

- 2014: https://blog.linkedin.com/2014/12/17/
the-25-hottest-skills-that-got-people-hired-
in-2014

- 2015: https://blog.linkedin.com/2016/01/12/
the-25-skills-that-can-get-you-hired-in-2016

- 2016: https://blog.linkedin.com/2016/10/20/
top-skills-2016-week-of-learning-linkedin

One of the other points I've learned during my career, which I try to reinforce in these books and in my classes, is the way that the core areas of digital marketing are related. For example, I consider search engine marketing to be tightly connected to all other areas in digital marketing. The central goal of digital marketing is to develop content that flows through various channels, including search engines.

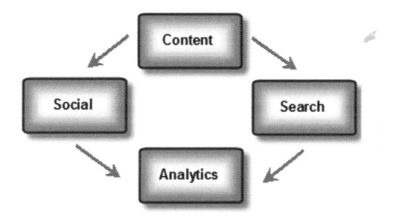

Just like each of the other areas, search engine marketing is crucial. If you do all the other steps, but you don't advertise, you won't get as much traffic, and it will be that much harder to succeed. On the other hand, as you'll learn, search marketing allows you to track return on investment, unlike almost every other kind of traditional advertising (radio, TV, billboards, etc.), and that's why it is so important. You can start out small, look at how much revenue resulted from the ads, and build with confidence. Search engine marketing has helped businesses around the world, both large and small, effectively sell on the Internet.

This book mentions what I call the core areas of digital marketing: Content, AdWords, Social, and Analytics (CASA). My goal is to reinforce how all the areas are connected. AdWords is Google's tool for creating ads for search engine marketing. The inspiration came from my professional background, as well as looking at trends in the marketplace.

The Core Areas of Digital Marketing

Content/SEO: search engine optimization is the process of attempting to boost your rank on Google so that you get higher up in search rankings when people type in particular keywords. Higher in search rankings = more clicks. The top way to boost rank is to add quality content that is relevant for your audience.

Adwords: the process of creating and managing ads on Google (Adwords), where you attempt to get people to click on your ads when they type particular keywords in Google. You pay when someone clicks.

Social Media Marketing: the process of creating and managing a presence on social media, including making posts, as well as creating advertisements. The main platforms are Facebook, Twitter, and YouTube, as well as Instagram and Pinterest

Analytics (Web visitors): You can gain valuable insights when you measure the performance of your websites and advertising campaigns. Google Analytics allows you to see how many people visit your site, where they come from and what they do.

Best wishes in learning search engine optimization!

CHAPTER 1

Introduction

I make liberal use of Google resources whenever I can, but sometimes it helps to curate them a bit. The quality and understandability varies; usually Google is technically accurate, but sometimes the results make more sense to an engineer, not a beginner.

I've come across a few videos that I recommend you watch, before you do anything else, and they present things dynamically and *interestingly*. It will help make these ideas concrete. Depending on how you are reading this book, you can go on YouTube and search for them there.

#1: How Search Works by Matt Cutts

If you end up pursuing digital marketing, you will probably want to seriously consider exploring Google's YouTube channel and look for blog posts and articles by Matt Cutts:

https://www.youtube.com/watch?v=BNHR6IQJGZs

#2: What Is AdWords?

Google AdWords is Google's tool for creating paid advertising. This book is about search engine optimization, which is also known as "organic" search, but it's helpful to understand how paid ads and unpaid organic results relate to each other, and this gives some context.

https://www.youtube.com/watch?v=O5we2g3Edgs

#3: Marketing in 60 Seconds: What Are Google Ads?

This is a non-Google video that I made with the help of a "whiteboard animator," who helped me to boil things down.

https://www.youtube.com/watch?v=lPsxn_lughQ

© Todd Kelsey 2017

T. Kelsey, *Introduction to Search Engine Optimization*, DOI 10.1007/978-1-4842-2851-7_1

How Search Works for You

I think it's helpful to point out that even though the inner workings of search can seem complex, when you're learning, you can always go back to the so-called "user experience". That is, what your own experience of a given tool or process is.

Without thinking about it, you are probably regularly engaged in the Google ecosystem of search engine optimization and search engine marketing. In fact, if you ever type in anything on Google, you're there.

For example, here is a slightly "older" view of Google, which I think serves our purposes:

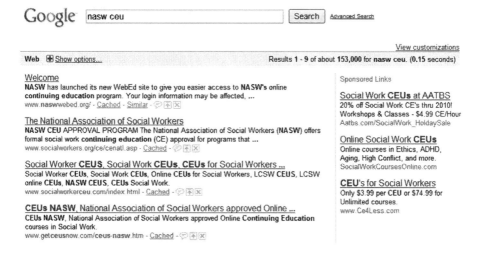

This is from around 2010, so the search page looks similar but some things have changed. If we run this same search today, it would look something like this:

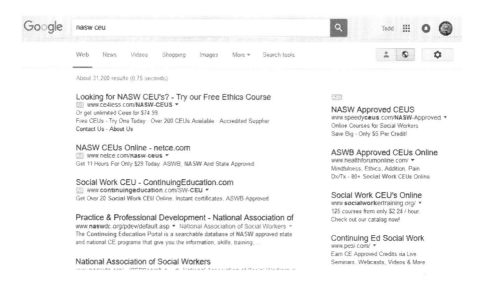

It looks a little more elegant, easier on the eyes, and has more whitespace.

The ads and "organic search results" are delineated a bit differently. In the next image, the ads, driven by the Google AdWords platform, are highlighted with boxes, and the arrow is pointing at the organic search result.

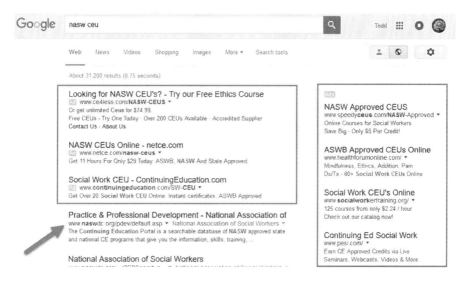

One way to think of it is kind of like a library, except Google is a for-profit library.

If you walk into a typical library, they'll have all kinds of free guides that list resources in your communities. But if they were a for-profit library, they sell advertising, and advertisers might compete with each other to get the prominent position for their ads. It's the same thing on Google—there's a space where you compete with other advertisers for paid ads that are driven by AdWords (see `www.google.com/adwords`).

But in a library, there are the free resources, and how are they typically organized? Probably neutrally—that is, using some kind of system that allows you to find what you're looking for, such as alphabetically, or some kind of searchable system. They might also *curate* things, such as organize them, or feature them, for browsing and exploring.

What Google, or any other search system for that matter, tries to do is guess what would be the most relevant result based on what you search for. Then, in the paid advertising space, companies compete to get the top spot, and in the unpaid, organic search results, Google will also list things based on relevance and popularity.

To get technical about it, there are several hundred known and theoretical ranking factors, and they affect "position" on Google. They determine how high your web site or article will rank on Google—the search engine optimization community is constantly trying to figure everything out. Google is public with some things and transparent with others.

But in the end, Google tries to keep things relevant and calculate a rank based on a variety of factors, including how popular material is. The more people who visit a site, the higher it will rank on a particular topic.

Google also takes into account other sites that link to yours, and the "authority" of those sites. For example, if you get a news article written about your site, media sources often have a lot of authority, because Google basically figures that if they cover something (depending on the site), it has a higher chance of being newsworthy. Since some legitimate human has made the choice to highlight it, Google pays more attention.

It All Comes Down to Quality Content

What I've seen in search engine optimization is that it has grown as an industry as Google has grown, and in the field of digital marketing, it seems to be the most constantly changing area. Best practices mean companies have to constantly react to what Google does, where they constantly seek to expand and improve how search results appear (and to impact their own profitability).

Naturally, marketing agencies, experts, workshops, and entire conferences and publications have arisen to address the area, and a great variety of techniques and philosophies and best practices have converged. A great number of people have also tried to *game* Google, by trying all kinds of tricks to get higher ranking in search results for their web sites and clients.

What I tell my students, and what I find constantly reinforced by the industry and by Google itself, is that in the end, it pretty much all comes down to quality of the content.

For example, if you did *nothing* other than write high quality, compelling, relevant articles on topics related to your business and organization, Google will find them, and more importantly, people will find them. If they're relevant and interesting, meaningful or helpful, more people will share them with other people. If this happens, they will climb higher in rankings.

I feel like the 90% figure is fair—if you spend 90% of your time on developing quality content (which could include video, of course), that's the biggest battle of search engine optimization. I've definitely witnessed efforts to master all the supposed "best practices" but which end up short changing the content itself.

Remember what I called the core areas of digital marketing? Content, AdWords, Social, and Analytics (CASA). The largest components of the field of digital marketing can basically be reduced to these parts—notice that it all begins with Content:

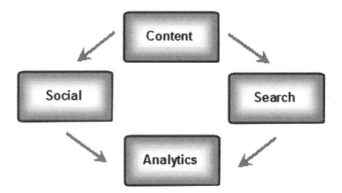

You need content to deploy on the social networks and messaging apps where people spend increasing amounts of time. For the time being, for the foreseeable future, people are using search engines to find information. What you want to determine is how did it all perform?

Not the Only Game in Town

This book focuses on Google, google.com in particular, but it's fair to say that it's not the only game in town. There are other search engines. As people's preferences shift and evolve, there will be other ways people search the Internet .

YouTube

For example, YouTube is technically, actually the second most popular search engine. People increasingly turn to it, not only for entertainment, but also to become more informed. This is especially true for people who prefer to learn from video rather than reading an article. Video content is definitely on the rise.

How does search work on YouTube? Similar principles apply.

For example, what follows is a silly music project I did in my past life in rock 'n' roll. The "real" band I was in didn't have a hit radio single, so I got together with some friends and made a joke, but we also made a few videos.

If you type the name of the project—"Gerbil Liberation Front"—into YouTube, at least at the moment, the top result is by someone who copied the video and uploaded it themselves. The reason it appears at the top is because they included the name of the project in the title of the video. (In "conventional" search engines, it's worth remembering that the title of an article is important.)

Whereas the second result (shown with an arrow) is a YouTube channel, so it doesn't rank as high:

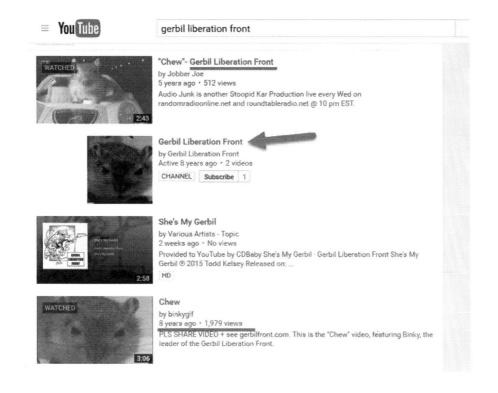

Then, very recently, the album was added to a site called CD Baby, which offers artists the ability to be included in a kind of automatic video that people can watch to get exposed to the music, so that shows up third.

All the way in fourth place is the original video upload, simply titled "Chew". In order to increase the rank, I could change the title to Gerbil Liberation Front. I could also try to write Google and ask them to remove the unauthorized content. Then, just as with any other search engine, I could try to get more traffic to the video—for example, by mentioning it in a book.

Bing (not Crosby)

Bing.com is another search engine, not as huge as Google, but still fairly popular. Techniques for getting higher-ranked content on Google apply to Bing as well, and in the end, it still comes down to quality content.

What's Next?

Not only does Google constantly evolve, but the entire industry of how people get information and media changes. For example, around the time of this writing, Apple introduced a new version of its operating system for iPhones, which has a feature that allows developers to more easily write "ad blockers". Guess what is the core of Google's 60 billion dollars a year in revenue? Ads.

If you consider how many people have iPhones out there (bajillions), then something like being able to remove ads could have a significant impact on Google. At the moment, pundits say various things—some tout doomsday, others say that not everyone will even be aware or run the ad blockers. Plus some ad blockers distinguish between "unobtrusive" ads like Google's. (Try searching on google.com—it appears really simple—then try visiting a page like cnn.com and notice the obtrusive ads.) Then there's evidence to suggest that Google might make deals with some of the ad blockers to let at least Google's ads through.

Apple is not the only game in town—the other half or so of the market for mobile phone operating systems is Android, which is a Google project.

But just sayin'—things change.

It wouldn't shock me if Apple and Google merged, or Google bought Facebook, or Google bought Twitter; but there's still a need for search engines, and I don't think they will go away. The search experience for some people probably started in libraries, looking for books on a computer system, or even a pre-computer card system. And then there is just browsing! Through books, video, music. Libraries have evolved and people don't visit traditional physical libraries as much as they used to. The search experience is there, and for most people, it begins on Google, YouTube, or Amazon.

Conclusion

Thanks for taking a look at this chapter, where we've scratched the surface of the nature of search, and considered where it's been, what it does, and where it's going.

In the next chapter, we'll take a look at an important toolbox for working with search engine optimization. You guessed it—Content!

Skillbox: Content

This chapter takes a look at the relationship between content related to search engine optimization, including tools you can use to do it yourself. You can think of it as a toolbox of skills, or a *skillbox*. We'll take a tour of some of the things I'm suggesting that you'll want to try when working with search engine optimization.

In some cases, if you're working for a company, they will already have systems and content, and may even have processes for search engine optimization. Still, there might be the occasional need to develop more, outside of the regular systems. During the learning phase, it can be helpful to know how to do it yourself, so you can get the feel of it or start straight out making "real" content as you're learning.

The goal is to introduce you to some concepts and tools that I think are worth trying, including for having material to work with as we learn more about search engine optimization through the rest of the book.

Curation vs. Creation vs. Collaboration

The general approach I recommend with content, especially if you're just learning about search engine optimization, is to think about each of these areas as connected.

© Todd Kelsey 2017
T. Kelsey, *Introduction to Search Engine Optimization*, DOI 10.1007/978-1-4842-2851-7_2

Curation

When you *curate* content, you go out and find it, and then share, review, and comment on it. You might just be gathering and collecting information. Even if you don't feel like you have a creative bone in your body, you can certainly go out and find material that might be interesting and relevant to your organization and then point to it and comment on it.

For example, in a blog post, you might find a collection of articles on a topic, write brief summaries/reviews, provide links to the original articles, and then offer some kind of conclusion. Or you might find YouTube videos that relate to a particular topic and link to them or embed them in a blog post. Then you might post your blog to social media, and thereby generate some traffic and awareness. The overall goal is to generate traffic to your site, based on providing relevant, compelling content, which will have a positive impact on search engine rank.

Creation

This is the best way to go if you can manage it, and can be the most fun. It takes more time, but it will probably result in the highest impact on search engine optimization. If the content is all you, you are less likely to lead someone to another web site. Ideally, you are inviting people to visit you, and not only giving them a reason to stay on your site longer than 60 seconds, but giving them reasons to *come back*.

Writing an article, posting to a blog, creating a video—all of these things are fair game, and if you go the extra mile to get know your customers and determine what they actually are interested in, you can create content tailored to the audience. What kind of content you create depends on the organization, but ask yourself who the audience is and consider what kinds of things they'd be interested in. You can ask people in person, by phone, by e-mail, on social media, and so on. It never hurts to test ideas and find out what people are interested in by going directly to them in some way.

Even if you aren't a "media professional," it can be helpful to develop some basic content. You can always ask others to review your efforts, and trying it out will help you be in a better position to source it. If you end up having the choice, it's better to focus on what you do best and hire someone to do the rest.

Collaboration

I think this one is helpful, especially for freelancers, independent business owners, and students. It can be a way to save money, a way to pool resources. You find people you can work with. It might be that you find someone who is a writer, offer to cowrite some content or do some research, and get the benefit of their writing skills or their reputation. If you want to make a video, you find people who are interested in the same outcome—raising awareness of a particular topic—or people who want to try something new just like yourself.

In other words, don't rule out anything. Even if you don't feel confident doing it yourself, can't afford to hire anyone, and don't know where to start, you might be able to find people who are looking to collaborate, even if it's just for the learning experience.

In a business sense, it can be the same thing. If a single local business owner doesn't have enough money for a particular project, you might be able to find a collection of business owners who want to pool resources for some kind of project (such as a video featuring local businesses). Or, using the Internet, you can find similar businesses in different areas and develop an article, a video, or content that could be repurposed for each business. With slight changes, the material could be re-used effectively.

Create a Google Account/Gmail Address

Google has a lot of free tools that make it easier to work with content, and when you have a Google account, it's easier to sign in to all the tools.

As a first step, I recommend creating a free Google account by going to `http://mail.google.com` and clicking Create Account. I advise my students in all my classes do this.

Content: Start a Blog (or Find One)

Blogs are an excellent starting places for developing content; they are basically specialized web sites for developing and posting articles on any topic you choose. A blog is a good example of how some of the core areas of digital marketing tie together.

As you saw in the first chapter, one way to think of the relationship of content to other areas in digital marketing is as a circle, where you start the whole thing with content:

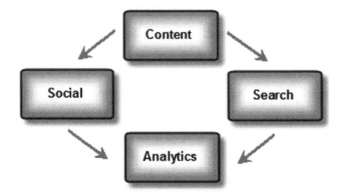

- *Content*: At a typical company, you might have a web site and possibly a blog as platforms for content. The idea is to draw people in and then *promote* the content.

- *Social/Search*: Then you take some of that content, a blog post for example, and do several things to promote it. You share that article on social media, and you run some ads to let people know about it. You're putting something out there, and the idea is to draw people back to your site. Google sees it and determines if the content is good enough. If so and if people click on it, Google gives you a bump in your search engine rank.

- *Analytics*: You can put all the effort into digital marketing you want, but if you're not measuring performance, what's the point? It's not just for show, it's to help your organization be sustainable. The best way to keep things clear is to measure performance, which means paying attention to analytics, such as the number of visitors to your web site.

If you want to start a blog, there are a variety of platforms for blogging, but blogger.com is one of the easiest to use.

My general experience with developing content is that the more complicated you get, the less sustainable it is. That is, you might try to use the most sophisticated platform, approach or tools, but in some cases that can slow you down, when all you really need is good writing and some visuals (or a video). I believe in using the simplest platforms possible, to help make it easier to *actually develop content*. Blogging and writing articles is one of those things that is easy to set aside or delay. The easier and simpler it is to make and post content, I believe the more likely it will be that you do it regularly. Simplicity is important. That's why I recommend Blogger as a starting point, and why I ask my students to create a blog as part of every class.

If you're interested in developing content as a skill, my general suggestion is to create a blog and set the goal of posting to it at least once a month. Choose a topic or tool you're learning about or a technique you're interested in, do some research, gather some links, and get in the habit of developing some ongoing posts. It will help keep your skills current and it will also be something you can point to when you're trying to get clients or find work.

Even if you already have a blog or one that's been untouched for a while, I still suggest creating a new one in Blogger. It's always helpful to learn new tools. Another reason it's helpful is because you could end up in a situation where a client might want to create a blog, and you can help them get started, by being familiar with and showing them different tools.

This same principle applies to some of the other tools we're looking at in this chapter. I recommend trying them in some way, both for own education but also because the skills could be useful for showing a client (or potential employer) someday.

So to get started, visit `http://www.blogger.com`. Either sign in with your Google account or click the Create an Account link at the bottom.

Google

One account. All of Google.

Sign in to continue to Blogger

Email

Password

Sign in

☑ Stay signed in   Need help?

Create an account

On the Blogger site, click the New Blog button.

For practice, I wouldn't be too concerned with the title. You can change it later easily, and you can also create/delete blogs easily. Feel free to use "Social Media Perspective" as a title if you want.

The title is simply what appears visually at the top of the blog. The Address field is the opportunity Google gives you to create a custom address. Because it's a free tool, you might have to experiment a bit until you find one that's available. Type your ideas in the Address field and see what happens:

What you're doing is coming up with the custom portion of the blog's address.

It turns out that, for this example, the address socialbuzznews.blogspot.com is available. The link for this blog is http://socialbuzznews.blogspot.com.

Then, after you choose a title and address, you can choose a template to determine the look and feel of the blog, which you can also change later:

Blogs List › Create a new blog

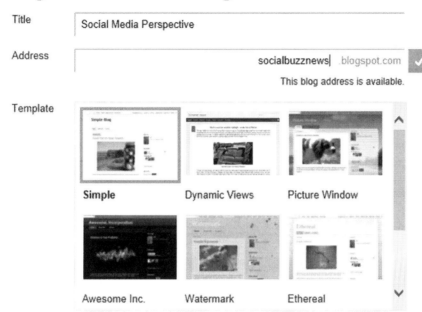

Title   Social Media Perspective

Address   socialbuzznews| .blogspot.com

This blog address is available.

Template

Simple   Dynamic Views   Picture Window

Awesome Inc.   Watermark   Ethereal

You can browse many more templates and customize your blog later.

Create blog!   Cancel

After you've selected a template (I recommend starting with Simple), click the Create Blog! button. Following these simple steps, you've created a blog and you can start blogging!

Your mission if you choose to accept it is to make a sample post, and then share the link on Facebook or via e-mail.

New Blog   Social Media Perspective   View blog

Your Blog has been created! Start posting | Dismiss

Note   One way to "cheat" if you forget the link for your blog is to click on the View Blog button (see the previous screenshot), which will open up the blog in your browser. Then you can copy the link from the address field and paste it into Facebook, or an e-mail, etc.

To learn more about Blogger, access the Settings menu (the little gear icon) when you're signed into Blogger and select Blogger Help:

There's a variety of helpful articles.

You can always go there directly using this link: https://support.google. com/blogger.

Weekly Bloggery: What I Do in My Classes

In case you picked the book up from Amazon, you might be interested to try the exercise that I require my students do in the classes I teach, which is simply to blog at least once a week. I often have them blog on specific chapters they are reading.

As a way to learn and as build up a habit of blogging, and as a way to impress recruiters (blogs do), I invite you to start blogging, on this chapter even, and to do so at least once a week. Go into Google Calendar and set up a weekly reminder if you need to. Do whatever it takes. It's a habit that makes a difference over time.

Simple suggested guidelines until it becomes a habit is to blog as little as 2-3 paragraphs and include a visual of some kind. Learn how to take a screenshot (I recommend using Snagit) or find a picture on a site like www. publicdomainpictures.net.

Just start blogging, simply, and describe something you're learning. To explore digital marketing more fully, go on over to social media (Facebook or LinkedIn) and share your blog post. Not only will you most likely have people who appreciate seeing it, it will also give you more confidence going forward. That's no big thing. You might think, "I'm blogging as part of learning how content impacts search engine optimization!". That sounds impressive, because it is!

Search Drill: Find a Blog

I recommend that you try creating a blog (and even setting a reminder to make a post once a month or more!).

I also a good idea to find a blog that you're interested in reading, as an example of "curating" content. You can find them using Google (search for "seo blogs"). Find articles that looks interesting and make a note of the links. Then take your readers on a tour of some of the blogs that you found helpful.

Content: Create a Free Web Site

Ultimately, when you talk about search engine optimization, you're talking about optimizing a web site. To review, a blog can be a web site on its own or it can be part of a larger web site, such as an e-commerce web site where you buy products. But in either case, whatever features a web site has (blog, e-commerce, etc.), it is primarily the content that results in people visiting, the site sending signals to Google, and Google rewarding you with higher search engine rankings.

Whether you are optimizing your own web site or working on someone else's, it can be helpful to consider creating an "official" web site for a project, event, campaign, or client. I like Blogger because it is simple to use, but it's not really designed to create general web sites. Other blogging platforms, such as WordPress (`www.wordpress.com`—there are free and paid versions), have grown to the point where they can serve as fully-functional web sites, depending on how you organize them.

If you're getting confused, just think of a blog as a place you post ongoing content, such as a library of articles, and your main site is the reference material that may not change as often, whether it's for a business or organization.

For general use, I think Google Sites is a good tool for anyone creating a simple web site, and more tools are mentioned at the end of this chapter. I recommend you try making a Google Site and keep it as part of your arsenal. You might even want to have a Google Site be your main site, such as your freelance business, with basic information.

To get started, go to http://sites.google.com. Log in if you need to (it's best to use a Gmail address created at mail.google.com) or click the Create Account button. (You do this to create a new Google account using your existing e-mail address.) You should see something like this:

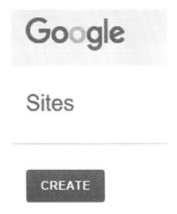

Like Blogger, Google Sites looks simple, but it packs a lot of power. The other advantage of keeping in mind sites like Blogger and Google Sites, is if you're not a developer, you don't need to have technical skills in order to create web sites using these tools. It might be a good alternative to provide the service of focusing on the content for the site or marketing, and whip up a site without necessarily having to hire a web developer. There are limits, of course, but it can also be a starting point. For example, use Google Sites to gather content, organize, and prototype, and then, when you have a better idea of where things are going, hire a designer/web developer or get costs and let your client choose.

I think the same applies to marketing microsites—say you develop a social media campaign and you want to have a microsite. A large part of the battle is developing the content—gathering it in the first place—and you could start with a free tool, take it to the limits, and then decide if you need a more flexible or professional looking design.

One other thing I'll say is that with the rise of mobile devices, the rules of design are changing a bit. It's not that you want to ignore design, it's just that you can't fit as many visuals on a smaller screen, and the mobile user is particularly interested in getting to the information. In other words, going to a site that looks okay on mobile devices (i.e., is simpler) is not such a bad idea.

So, back to Google Sites.

It's that easy. You click the Create button. (Note, in this chapter, we use the Classic Sites option; depending on when you read this you may have an option for the New Sites).

If you're taking it for granted how easy it is, try going to Godaddy.com and look at how much effort is required to start a hosting account and get a web site started. Whether you're using a content management system like Drupal, which is a website builder or using a manual tool like Dreamweaver, it's a lot of work. I guarantee after trying that, you'll appreciate how much time you're saving by just being able to click Create. Thanks, Google.

As with Blogger, there are some prebuilt templates you can choose from. In It's a little bit trickier to go back and change things in Google Sites, so until you spend some time really exploring it, I recommend choosing the Blank template at first.

You select the template and then choose an address, just like you do in Blogger.

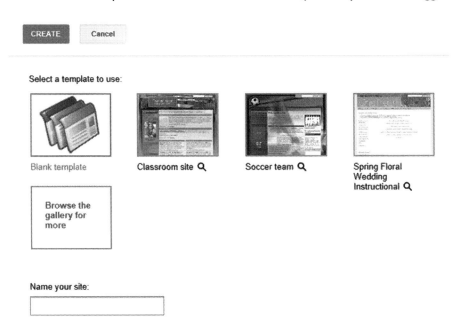

You can click in the Name Your Site field and type in a name, which is like a title, and can be changed easily later. Then you have to experiment and try different "site location" names. You can also click on Select a Theme, or on More Options, but to begin with I suggest keeping it simple. (In Google Sites, you can always come back and change the theme later. Themes provide the basic look and feel of the site.)

Name your site:

socialmicrosite

Site location - URLs can only use the following characters: A-Z,a-z,0-9

https://sites.google.com/site/   socialmicrosite

▸ Select a theme

▸ More options

Type the code shown:

CREATE

You also have to type in the CAPTCHA code (such as "bitsu" shown previously) before you click the Create button.

As you're typing in site location names, Google may tell you that the one you want isn't available:

Site location - URLs can only use the following characters: A-Z,a-z,0-9

https://sites.google.com/site/   mysite   ✕

The location you have chosen is not available. Learn more...

Keep trying until you find one that's available. Voila! You have a new web site.

To learn more about Google Sites, go to either of these links, which point to the same place: `https://support.google.com/sites/?hl=en#topic=1689606` or `http://tinyurl.com/googsite-help`.

My recommendation is to begin by making a basic site. It could be for yourself or your portfolio, or for a potential client, such as an imaginary local business or a promotional campaign of some kind.

■ **Note** There is a "classic" version of Google Sites, and a new one that Google is rolling out. I recommend trying the classic version first, until the new one works. Google may give you a choice of which to use.

Other Systems for Making Web Sites

A couple more options that are popular and have free/paid options for making web site, include:

- `http://wix.com`
- `http://weebly.com`

Take a look and try them out. Even if you're an intern at a large corporation, sometimes big companies use microsites too—especially in marketing situations.

Personally, I have been pretty impressed with Weebly.

Make/Edit a Video (or Find One)

It's true that entire books, or sets of books, could be written about each individual section in this chapter, but I'm recommending some simple starting points that are helpful for making content. With video, as with blogs and web sites, you may end up wanting to hire an agency or professional to do the final video. At the same time, individual can do pretty interesting things on their own, with something as simple as an iPhone.

There are a variety of ways to make videos, and if you don't have a video camera or iPhone, I recommend getting an older iPhone or an iPod (yes, they still make them) for the purpose of filming video. You can get a used iPhone fairly cheaply, connect it to WiFi, and upload videos directly. As of this writing, even a new iPod is $200, which is not too bad, with no monthly cell phone bill.

(It's also true that a used iPhone could be used for making video and some other uses, even over WiFi.) If you've always wanted an iPad, there are used options there too.

The reason I recommend one of these options for experimenting with video is because it's simple, easy to use, creates high-quality video, and you can upload the video directly to YouTube.

Getting familiar with it, including making simple videos (such as interviewing someone) and posting them on your blog or web site, can be good experiments—all good ways to build skills and your portfolio. In more professional environments, even if you're requesting a budget to hire a professional videographer, you might still want to try prototyping your ideas with cheaper modes of video.

Keep in mind that some of the most popular videos on YouTube were created with really simple equipment—it's more about the idea than the equipment.

I'm not Apple-biased, it's just simple, and that's good for beginners. Point and click and you can make simple videos. Then, you're one click away from being able to upload to YouTube:

Part of the reason I am suggesting this kind of arrangement is to keep the technical hassles to a minimum, so you focus on the content. The easier it is, the more fun it is, the fewer hassles there are, and the more confidence you will build from trying things out. Just a starting point.

By comparison, you can certainly get a digital camera with built-in video capability, or any number of dedicated cameras, load video editing software on your computer, and then transfer the video to your computer. In fact, I encourage you to explore that approach at some point.

But to begin with, I recommend getting a cheap iPhone or iPod or iPad, perhaps one that isn't the latest version. Just make sure it has built-in video.

I'm suggesting you try some videos without even editing or just by using YouTube's built-in editor. Meaning you shoot some short clips on your simple mobile device and upload them to YouTube. Then you edit them on YouTube. Simple, fairly easy, and lets you focus on the content.

Another option is certainly Android, an alternative to an iPod or iPhone. Another option is the Android, an alternative to an iPod or iPhone. My general experience is that Apple takes a lot of time to provide a good, simple, fairly stable user experience, whereas with the Android platform, it depends on the manufacturer. The time you spend on figuring things out may take away from having fun. However, to save money, or to avoid Apple specifically, Android tablets are certainly a good option.

At the time of this writing, you can get an Android tablet, with the capability to shoot video and upload it to YouTube, for about $50 USD.

Editing on YouTube

My recommendation is not to worry too much about the video; just try shooting "something," even if it's really simple, and then upload it to YouTube.

The YouTube Editor is located at `https://www.youtube.com/editor`.

Even though you can do some very basic editing on a phone, you don't necessarily need video editing software on your computer. On a Mac you're set for starters, they come with iMovie. Some Windows-based machines include basic video editing software as well. The point is that now you can upload video to YouTube and edit there. Why would you want to edit? To focus things, to pick out clips you want, reassemble them, add some music, and so on.

Getting Going on YouTube

If you are new to Google accounts or to YouTube, you might need to set up a profile on YouTube:

How you'll appear

Add photo	John	Doe

To use a business or other name, click here.

Gender

Male ▾

Birthday ⓘ

April ▾   1 ▾

Cancel   Continue

Then, the YouTube videos you've uploaded to your account will appear there, and the YouTube Editor allows you to do some basic editing online. I think it's nice, even if there are limits, to be able to try video by shooting on a mobile device, uploading directly, and then editing right there online.

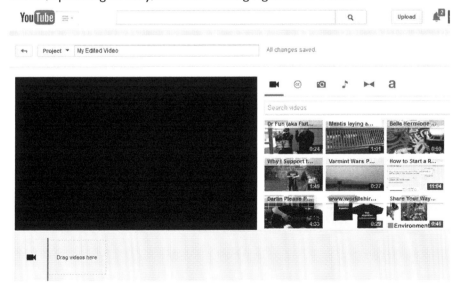

As with the other topics, entire books could be written about on shooting and editing video. You may want to look at some, but I also encourage you to try an experiment of making a simple short video and not worrying too much about technique yet. I think it will give you confidence.

With regards to developing content to help boost search engine rank, it's worth noting that YouTube is the second most popular search engine.

That's right, YouTube is a search engine—for videos. So you should develop video and post it on YouTube, and then actually embed it in your blog posts or on your site. (See the front page of `www.rgbexchange.org` for two "embedded" YouTube videos.)

Putting videos on YouTube is also a form of search engine optimization, only your goal on YouTube itself is to get as many people to come to a section of the site, based on getting ratings, etc. Technically, though, it is still a web site—a section of YouTube that you can manage, where you post your videos and put other information.

YouTube is also a social network, so it's part of content development and social media marketing. I think video is probably one of the strongest, long-term things that will still be around as social networks rise and fall. Getting familiar with making a video, even if it is rough or simple or mainly informational, is a good thing.

More Reading on Video Editing

Here are some links to explore in order to help you get started. (I invite you to consider reading each article, writing a few sentences about it, making that one of your blog posts, and then sharing it! You'll be glad you did).

- Use YouTube Video Editor: `https://support.google.com/youtube/answer/183851?hl=en` or `http://tinyurl.com/useyteditor`

- Quick and Basic Video Editing with the YouTube Editor: `http://www.howtogeek.com/howto/23346/quick-and-basic-video-editing-with-the-youtube-editor/` or `http://tinyurl.com/ytquickandbasic`

- Basics of Video Editing: `http://lifehacker.com/5785558/the-basics-of-video-editing-the-complete-guide` or `http://tinyurl.com/lhvideoediting`

- Introduction to Basic Concepts: `http://diyvideoeditor.com/video-editing-basics/` or `http://tinyurl.com/diyvideobasics`

You can also go right to the source. To learn more about the YouTube Editor and how to use YouTube, go to `https://support.google.com/youtube`.

Welcome to the YouTube Help Center

Getting started on YouTube

How to navigate YouTube

Build playlists of videos you like

Subscribe to the channels you love

Watch videos and playlists

Create videos and manage channels

YouTube accounts

YouTube Partner Program, including monetization

Paid content on YouTube

Copyright and rights management

Policies, safety, and reporting

Advertising opportunities on YouTube

Adjusting Digital Images

Another area I recommend exploring for the purposes of social media marketing is digital images. This can mean something as simple as looking on Google for images. Once you find an image you like, you can right-click on it and save it to your computer (Windows) or hold Ctrl down and click on the image to copy it (Mac). Try some simple image searches and practice with those images first. Then explore royalty-free image collections and services like `clipart.com`.

The point is, eventually you'll probably want to at least try playing with adding images to a blog post or web site, and it's good to know how to work with images or pictures for posting to social media. For example, even when you have a text-based post, an interesting, related picture can lead to more attention.

Even though when you upload images to web sites or blogs, there are often built-in "resizing" functions, you still might want to try a tool like `www.picresize.com` to crop an image or resize it. (Try looking at the help section.)

In other words, there are online tools you can use, without necessarily having to learn a complex image editing program like Photoshop (costly) or Gimp (open source at `www.gimp.org`).

Taking Screenshots

For grins, and to add some value to your blog posts, I recommend learning how to take screenshots, especially students and interns—or anyone for that matter. Think of it as a way of taking pictures on the web. For example, most of the images in this book are screenshots, where I take a picture of something and then discuss it.

One reason you might want to do this is because in a blog, you might want to add visuals, and taking a picture of a piece of software, or a web site, is an easy way to get a visual.

Greenshot is one free tool you might like to try. You find it at www. getgreenshot.org.

On the Mac, you can also press Cmd+Shift+4 on your keyboard to capture whatever is on the screen. (See http://guides.macrumors.com/Taking_Screenshots_in_Mac_OS_X for more information.)

Your mission should you choose to accept it is to choose a feature in something like Blogger, Google Sites, or even Facebook, explore it, take a screenshot or two of it, and put it in a blog post.

I also recommend taking a serious look at Snagit. There's a free trial, and it's a good way to be able to add arrows and text, easily edit images, take screenshots, etc. (it's what I use for the books I write).

Conclusion

Congratulations on getting your feet wet with what I believe is the true foundation of search engine optimization—actually making content. Google is always working on updating things, with the goal of making it easier for you to focus on what's most important—the content, as opposed to all the technical things that go around it. Blogger and Google Sites are good examples of this. They are excellent learning tools for non-profits or small companies that need to be frugal. Then there are always the slightly more sophisticated options, such as WordPress, Wix, and Weebly—the three Ws!

Try one new product or tool each week, and of course blog about it. Use Blogger, Google Sites, Wix, Weebly, or WordPress. Go for it.

Best wishes getting started with content!

▓ **Note** What are you supposed to write about? Search engine optimization, of course! You don't need to be an expert. Just explore as if you were touring a different country, take pictures on your travels (screenshots), and then tell people about your experience in blog posts.

SEO Basics

This chapter looks at some SEO basics, including on-page and off-page optimization, as well as related concepts and questions you should ask.

SEO: Where's the Content Going?

First of all, it's important to ask where the content goes:

- Social media

- Ad campaigns

Ultimately, that content should be on *your* web site. The principle of search engine optimization is to draw people to *your* site with good content. Think of your web site like real estate. More content means more visitors, which leads to more Google notices. More Google notices lead to a higher rank, which translates to higher value.

The important point is to start thinking about is how content can be created and deployed on your web site. One simple place to start is to provide *informational* content that helps viewers learn about your products, as well as generic, related topics. Part of the idea in developing content to boost search engine rankings is that you want to understand your customers well enough (by talking to them) to understand what they are interested in learning about, *even if they are not ready to buy.*

Some statistics suggest that of the people who visit your web site, only 2% are ready to buy or take some kind of action. The other 98% may just be doing research, exploring possibilities, and so on. With this in mind, it may be that

© Todd Kelsey 2017
T. Kelsey, *Introduction to Search Engine Optimization*, DOI 10.1007/978-1-4842-2851-7_3

the learning content on your site alone is what draws them in. By providing information about a relevant topic for them to read, you draw them in. For example, a whitepaper, user guide, blog post, etc.

Content SEO

This is probably not an official term, but I'm going to use it any way. What we're talking about in this section is *Content SEO*.

To simplify even further, here are the basic steps for Content SEO:

1. Add more content
2. Get more visitors
3. Get more Google notices
4. Get a higher rank on Google
5. Web site increases in value

Other search engines follow a similar process—Bing, Yahoo, etc. (Those are the main ones, but Google is the overwhelming leader in most of the world. In China, Alibaba reigns supreme, and other countries have unique leads in certain cases, like Naver in South Korea).

Here is my top secret graphic that summarizes the entire book and field of SEO. If you remember a single thing about SEO, remember this:

IMHO SEO 90/10 Rule

90% of SEO is content

10% is technical tweaks

IMHO means in my high and exalted opinion (okay, I mean in my humble opinion).

Get it? It's about the content, and only 10% is about technical tweaks. In other words, if you just put good quality, relevant content up on your site and never do another thing with SEO, that would be something like 90% of the battle. Conversely—and this is the important thing—if you spent your time learning about and mastering all the latest best practices of technical SEO, such as the latest changes in algorithms or fine-grained coding techniques, it might be helpful, but if it's not built on a foundation of content, it's not going to be very impactful. It'll do some good, but not as much as if you have good content.

Classic SEO: On-Page, Off-Page

Let's start to get our feet wet with the 10% of SEO that involves technical tweaks, what's called "classic SEO". It's the kind of thing that companies and organizations see as increasingly valuable, and for good reason. It's important enough that SEO is right up there with SEM (search engine marketing), and is a top skill out of any skill that get people hired (based on LinkedIn's annual "top 25 skills that get people hired).

As long as you remember the 90/10 rule—content is the most important thing—you're good.

One good way to understand how it works is to think about what they often call an "SEO audit". So as a new employee at a company that's never done SEO before, or as a team member at an agency that is helping a new customer, you might look at several factors as part of an SEO audit:

- *See*: Can Google can "see" your site properly
- *Here*: On-page factors
- *There*: Off-page factors

See: When you post a web site on the Internet, eventually and probably pretty quickly, Google will find and scan it. This is also known as *crawling*. Google goes to great lengths every year to improve the automatic way they scan and classify the Internet, but you can make it easier, including making sure all of your content can be accessed easily. Why make it hard, right?

Here: On-page SEO involves things you can do on your own web site to enhance search engine rank and to make sure your web site/page is set up correctly. You could also think of this as "on-site" SEO.

There: So called "off-page" SEO are tactics and considerations for thinking about how you get traffic from other web sites. One of the classic techniques is to think about links—getting links, adjusting links, deciding how the link is going to work, and so on.

SEE, HERE, THERE. Okay, that's it, the book is done. You've got it!

Just kidding. I tend to oversimplify things a bit. The reason that I highlight that content is 90% of the battle is because in some cases the so-called "technical tweaks" have gotten out of hand. This includes entire conventions, companies, and individual consultants. Some of them have made wild claims about being able to guarantee getting a particular search rank, or getting it on the first page of search results. In some cases on the shadier side of things, some companies would try to play games with Google, and this is something you definitely do not want to do. If it sounds too good, whether it's a "best practice," or if it's a paid service, you need to understand what techniques a company is using, and if they don't really talk about it, then you could (and should) say, "We don't want any black hat SEO".

Black hat SEO involves playing games with Google, which can get you banned. Sometimes it's grey hat SEO, and the only way to learn about it is to look at it a bit.

Keywords and More Keywords: Can you Ever Have Too Many Keywords?

At this point, it is fair to stop and think about keywords for a moment. Keywords are at the heart of SEO. If you haven't watched the video titled, "How Search Works" that was mentioned in the last chapter, you should put the book down right now and watch it.

An easy way to think about keywords is that people have to type something in when they search, right? You could think of a keyword like a fishing pole.

search results

If you type in a keyword or phrase, it's kind of like using a fishing pole, where you throw it out in the water and hopefully catch something. On Google, when you type in a keyword, you get *search results*, not fish.

If you were to use your fishing pole (Google) and type in the phrase "non-profit stock exchange" you might get something that looks like this:

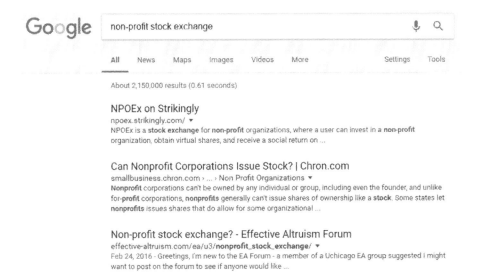

The important point to remember is that search engine optimization is ultimately based on keywords. That is, what are the phrases that people type that will lead them to your site?

Examples of common keywords include:

- The name of your company or organization
- The name of a product
- The generic category of your product

If you typed in "RGB Exchange", you might get something like this. The RGB web site comes up as the first search result. Woo-hoo!

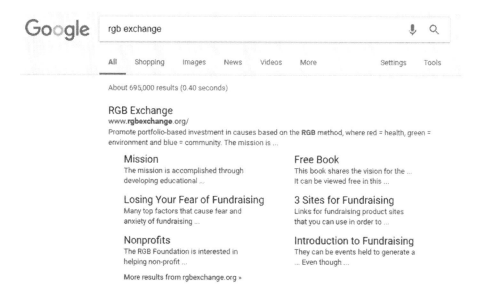

This is sometimes called a "brand" keyword, and it's a little easier for such keywords to get top rank than others, because there is less competition.

Another common search keyword involves either a generic product name or a specific brand name product. Let's try the generic product type. Say you are a safety engineer and have electrical equipment of all kinds. As part of your industry's safety standards, you have what's called a lockout tagout station, which is a system that ensures that the right switches stay turned off when you are working on electrical equipment, so that no one gets hurt.

You could try typing "eco-friendly lockout tagout station" in Google, and you might end up with something like this:

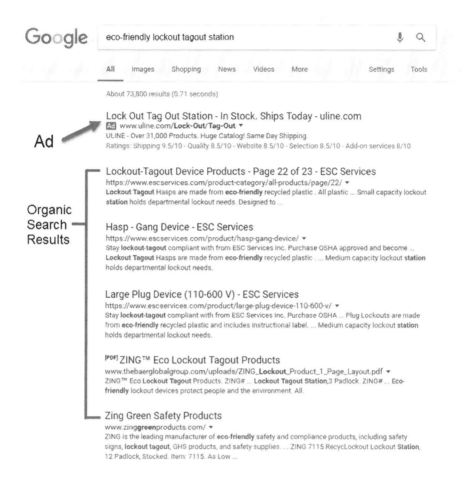

Note the ads and the organic search results. In this scenario, the Zing Green Safety Products site comes up as a search result, based on the keyword "eco-friendly lockout tagout station". The rank for the keyword is good; it's on the first page.

To understand the relationship between SEM and SEO, you can see that there is an ad at the top that appears when the keyword is typed in. If someone clicks on it, the company will *pay* for that click. For the subsequent search results, the "organic" search results, no one pays for the click—technically it's free. Except that companies spend a lot of money to do SEO, which is the process of trying to get your search result as high as possible, ideally at the top of the first page—for whatever keyword or keywords you are going for.

If you find yourself getting confused or intimidated, remember the marvelous 90/10 rule. Even if you just post content to a web site—quality compelling content—you've fought 90% of the battle. From that standpoint, you don't even need to think about keywords. Some SEO professionals might balk at that claim, but evidence increases every year that Google rewards quality content over just about everything.

You could spend time editing an article, for example, based on emphasizing particular keywords, but if you try too much of that, it might sound unnatural, right? So technically you might have followed a typical practice of pro-actively inserting and emphasizing, and even repeating, particular keywords, because those keywords are on the "hit list" for the web site. If you do this too much, or if it diminishes the underlying quality of the text, you're just going in circles, because the content won't be as compelling.

What I generally recommend then, for learning about "technical tweaks" and as you start to be aware of keywords, is simply to be aware of them, to consider investing time in some keyword-level optimization, but definitely place a priority on quality content.

In the previous example, a goal of SEO optimization for Zing Safety Products might not only be to maintain the first page search rank, but to try to increase the rank for that keyword, to get the organic search result as high on the page as possible. Why? Because it's like free advertising, in a way.

Okay, so how do you do it? You could develop some content on the site, related to that keyword, such as an introductory article (or blog post) that describes it. Make it relevant and helpful enough, and not only will people have a better view of your company, but they might also share it with others, and so on.

To return to the "game playing with Google," basically most of that involves trying to determine exactly how Google works, and then trying techniques to get as high a search result as possible, often using unnatural techniques, like taking keyword insertion and going to extremes. An example of this in earlier black hat SEO was to include invisible keywords repeated over and over on page. They were visible to user, but were visible to Google.

Keyword based SEO, such as optimizing, adding, inserting, and managing, is an example of on-page SEO. It's something you do on your own web site.

Off-page SEO is something like trying to get a different site to link to your web site. Along with many other ranking factors, some known, some revealed by Google, others only guessed at, "inbound links" to your web site can contribute to rankings, but they are not on your site.

Like when using keywords, there is a natural way to do it, and you can get unnatural as well.

For example, if you write compelling content, and if it is compelling enough that people share it on social media—that is like a link back to your site. It may very well produce traffic, and Google notices things like that. It may be that a product review web site has an article, along with a link to your site. That's another example of natural off-page SEO; it happens because of the quality of your content or product, not because you proactively sought the link. You can take one step further and try to find places to review your product or site—that's pretty natural. Another approach might be to get coverage by a blogger, news site, etc.—this is also fairly natural.

But if you go too far, such as participate in some kind of "sounds too good to be true" link-building scheme, where a company promises you the world for a cheap price, eventually Google will penalize you for these black hat practices.

So just go natural, okay?

Example: SEO Steps for a New Web Site or Client

To return to a web site SEO audit, there are several things you can look at. I simplified it to see, here, and there.

- *See*: Be sure that Google can "see" your site properly
- *Here*: On-page factors
- *There*: Off-page factors

There are a variety of approaches and philosophies about how to analyze a web site initially, and because things change fairly regularly at Google, it's a good idea to learn the latest techniques. Sites like Search Engine News and Moz are good sources of ongoing information. The latest information can make a significant difference, when you get down to the technicalities. Check them both out:

- www.searchenginenews.com
- www.moz.com

For example moz.com has a good article, entitled "How to Perform the World's Best SEO Audit". Check it out at:

https://moz.com/blog/how-to-perform-the-worlds-greatest-seo-audit or http://tinyurl.com/mozaudit

I invite you to read it and then to consider what just happened:

1) I searched on Google for articles on SEO audits, and theirs showed up toward the top of the list, because of how good and relevant it is.

2) I found it relevant and helpful enough to pass along to you, my readers.

3) Some of you will click on the link in the electronic version of the book and go right to the site; others may type in the really long version of the link to the article and go to the article. Google will notice this activity, and it will protect and strengthen the "SEO rank" for the article.

So voila! It's a perfect example of content SEO. But read the article and see what it has to say. Another article that I like has a simplified series of steps to help you understand and learn what kinds of technical tweaks are effective during an SEO audit. For example, see First 9 SEO tasks at http://www.plusyourbusiness.com/first-9-seo-tasks/ or http://tinyurl.com/9seotasks.

Read what they have to say and then compare the two articles. There are some similarities and some differences that might be a result of either different philosophies or being written at different points in time.

This might be the second most important point to take away from this chapter—regularly review resources like www.searchenginenews.com and www.moz.com. The only thing that doesn't change is the fact that things always change. In other words, on a regular basis, Google makes changes to its algorithms, which changes the way things rank.

Activity: Meta Description

Before we get too much into what *meta* means, I have to mention HTML. You may have heard the acronym, it stands for HyperText Markup Language. It's like a screenplay language where the code is the director and it tells the actors (the browsers) what to do. It's like a screenplay for a browser, telling the browser what to do.

If you've never seen HTML before, right-click (in Windows) or hold the Ctrl key down and click (on the Mac) on a browser window and choose View Source (or View Page Source—depends on your browser). You can then see what's going on behind the scenes. As an example, feel free to do so on the www.rgbexchange.org page.

You might see something like this:

```
<!DOCTYPE html PUBLIC "-//W3C//DTD XHTML 1.0 Transitional//EN" "http://www.w3.org/TR/xhtml1
<html xmlns="http://www.w3.org/1999/xhtml" itemscope="" itemtype="http://schema.org/WebPage
<head>
<meta http-equiv="X-UA-Compatible" content="chrome=1" />
<script type="text/javascript">/* Copyright 2008 Google. */ (function() { (function(){funct
{this.t[a]=[void 0!=b?b:(new Date).getTime(),c];if(void 0==b)try{window.console.timeStamp("
a,d;window.performance&&(d=(a=window.performance.timing)&&a.responseStart);var f=0<d?new e(
c=a.navigationStart;0<c&&d>=c&&(window.jstiming.srt=d-c)}if(a){var b=window.jstiming.load;0
0,c),b.tick("wtsrt_","_wtsrt",d),
b.tick("tbsd_","wtsrt_"))}try{a=null,window.chrome&&window.chrome.csi&&(a=Math.floor(window
0,window.chrome.csi().startE),b.tick("tbnd_","_tbnd",c))),null==a&&window.gtbExternal&&(a=w
(a=window.external.pageT,b&&0<c&&(b.tick("_tbnd",void 0,window.external.startE),b.tick("tbn
{}})();  })()
</script>
<link rel="shortcut icon" type="image/x-icon" href="//www.google.com/images/icons/product/s
<link rel="apple-touch-icon" href="http://www.gstatic.com/sites/p/662db1/system/app/images/
<script type="text/javascript">/* Copyright 2008 Google. */ (function() { function d(a){ret
g(a){return a.replace(/^\s+|\s+$/g,"")}window.trim=g;var h=[],k=0;window.JOT_addListener=fu
{eventName:a,handler:b,compId:c,key:f};h.push(a);return f};window.JOT_removeListenerByKey=f
{h.splice(b,1);break}};window.JOT_removeAllListenersForName=function(a){for(var b=0;b<h.len
window.JOT_postEvent=function(a,b,c){var f={eventName:a,eventSrc:b||{},payload:c||
{}};if(window.JOT_fullyLoaded)for(b=h.length,c=0;c<b&&c<h.length;c++){var e=h[c];e&&e.event
(e="function"==typeof e.handler?e.handler:window[e.handler])&&e(f))}else
window.JOT_delayedEvents.push({eventName:a,eventSrc:b,payload:c})};window.JOT_delayedEvents
```

Ack!

Don't be alarmed. Even though a lot of the code looks like gobbledygook and is not necessarily legible to the human eye, some of it is. For example, with the View Source window open, you could press Ctrl+F, to search within that window, and type the word "meta":

```
1  <!DOCTYPE html PUBLIC
   "http://www.w3.org/TR            meta          1 of 10   ∧  ∨  ✕
2  <html xmlns="http://www.w3.org/1999/xhtml itemscope
   itemtype="http://schema.org/WebPage">
3  <head>
4  <meta http-equiv="X-UA-Compatible" content="chrome=1" />
5  <script type="text/javascript">/* Copyright 2008 Google. */
   (function() { (function(){function e(a){this.t=
```

Then click the down arrow until you reach something like this, which is what we're searching for:

```
<link rel="canonical" href="http://www.rgbexchange.org/home" />
<meta name="title" content="RGB Exchange" />
<meta itemprop="name" content="RGB Exchange" />
<meta property="og:title" content="RGB Exchange" />
<meta name="description" content="Promote portfolio-based investment in causes based on the RGB method, where red =
health, green = environment and blue = community. The mission is accomplished through developing educational
materials, and providing secondary services such as pro-bono assistance and consulting. The foundation is also
developing a pilot for a donation platform that will provide an investment experience.&#xA;" />
<meta itemprop="description" content="Promote portfolio-based investment in causes based on the RGB method, where
red = health, green = environment and blue = community. The mission is accomplished through developing educational
materials, and providing secondary services such as pro-bono assistance and consulting. The foundation is also
developing a pilot for a donation platform that will provide an investment experience.&#xA;" />
<meta id="meta-tag-description" property="og:description" content="Promote portfolio-based investment in causes
based on the RGB method, where red = health, green = environment and blue = community. The mission is accomplished
through developing educational materials, and providing secondary services such as pro-bono assistance and
consulting. The foundation is also developing a pilot for a donation platform that will provide an investment
experience.&#xA;" />
<style type="text/css">
```

The underlines in the graphic highlight the "meta tags," which are part of HTML code. These particular tags send Google signals for search engine results. In some cases, web platform systems automatically generate these tags, guessing what you want the description to be. In the case of this page, what is contained in the meta tag affects the search result:

Notice the phrase "promote portfolio-based investment" in the code example and notice how that description is what appears under the search result. It's just the meta tag, telling Google what to do.

For technical tweaks, create a free web site at weebly.com and then go into the Settings area to look for SEO.

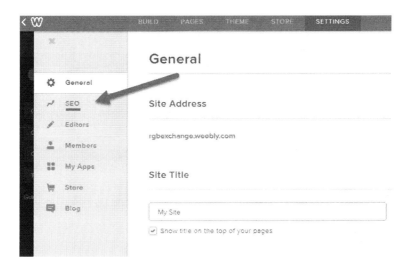

Then, enter some text for a description for your site in the site description area.

Without necessarily knowing the HTML code, you've done a technical tweak, and it has had a direct impact on the technical code. For extra points, publish your web site, then use the View Source technique to search for meta. See if you can find what you entered as the SEO description, within the meta tag. Okay, go!

Here are a few additional related articles that are worth reading:

- `http://moz.com/learn/seo/meta-description` or `http://tinyurl.com/mozmeta`

- `https://support.google.com/blogger/answer/` `2472665?hl=en` or `http://tinyurl.com/googblogseo`

The latter article talks about how you can do some basic SEO optimization on a Blogger blog.

To explore HTML a bit, to get in touch with your inner HTML, try this friendly short article: `http://www.casamarketing.org/seo-and-html`.

Conclusion

Congratulations on making it to the end of this chapter—this was your first dry run through some technical areas. Just remember the 90/10 rule (content is 90% of the battle!) and you'll do fine.

Try to have some fun playing with the content tools, like weebly. You can create a free, quick web site, lickety split—no problem!

Keyword Research

Keyword research is the first step in SEO and a fundamental best practice. At some point it's a good idea to start thinking about keywords, and it can also be a part of an SEO audit.

If you end up in a situation where a company or small business already has SEO operations, they might have identified keywords that they are targeting, as part of their SEO or search marketing efforts (where they are jumping above organic search results and buying ads on Google). Even if there are some insights available, it can still help to regularly look for new opportunities for keywords that you can target with both SEO and SEM (search engine marketing). Understanding the relationship among SEO, SEM, and keywords can also help you ask the right questions. For example, a company might have a person working on SEM (i.e., buying ads on Google and other search engines), and that person might have insights on keywords to target, even if they weren't doing SEO before.

So this chapter takes a closer look at keywords, the way SEO and SEM relate, and some of the tools you can use to explore keywords.

Understanding Keywords

If you're not quite sure yet how keywords work or how they relate to SEO, the best way to get familiar with the process is to go to Google and search for a product. Let's use tennis rackets as an example.

© Todd Kelsey 2017

T. Kelsey, *Introduction to Search Engine Optimization*, DOI 10.1007/978-1-4842-2851-7_4

Remember to determine whether a search result is an ad or an organic search result.

Top 10 Tennis Rackets - Best Tennis Rackets Of 2016's
[Ad] www.bestreviews.guide/**Tennis-Rackets** ▾
We Did The Research For You.

Tennis Racquets | DICK'S Sporting Goods
www.dickssportinggoods.com/products/**tennis-racquets**.jsp ▾
Get in the game with a new **tennis racquet** from DICK'S Sporting Goods. Shop a variety of **tennis racquets** for juniors & adults from top-rated brands.
Adult Tennis Racquets · Tennis Racquets For Kids · Unstrung Racquets

In this scenario, I typed in *tennis rackets,* and an ad was displayed for www. bestreviews.guide. The thing to remember is that the ad appeared because the company *targeted* particular keywords. That means that over time, they asked questions like this:

Q: What keywords are people using to get to our site?

A: In some cases you can know this, by using Google Analytics on your site (see *Introduction to Google Analytics*) and connecting your site to Google Webmaster Tools. In Google Analytics, you look at the keywords that people use to get to your site, through Google organic searches (meaning not through an ad). This might result in keywords that you already knew about, but it might also reveal other keywords.

Q: What kinds of keywords are people who *might* be interested in your site using?

A: To me, this is the central question of SEO and SEM, whether you are going for organic search results or working on choosing keywords for an ad campaign. This is where you try to imagine current customers or supporters, or potential customers. You try to imagine the kind of person who might be interested in what you have to offer, and then target the kind of keywords they use.

Keyword Opportunities

Those keywords out there that people are typing in represent an opportunity. As you work through the overall process of SEO, you should develop this keyword mindset.

For example, let's say you work for a company that makes drone kits. Then imagine that John Doe is interested in building a drone. What kinds of keywords might he type into Google when he's doing research?

Try imagining some:

- How to build a drone
- How do I build a drone?
- Drone kits
- Best drone kit

These are generic phrases, but John might also search for brand name phrases:

- DJI drone kits
- 3DR drone kits
- Erle copter drone kit

He might also search for a very specific type of drone kit or part, such as:

- Carbon fiber drone frame
- Carbon fiber racing drone frame

Don't worry if you don't know anything about drones (I don't). Just remember that keywords can be pretty generic, but you can also get very specific and detailed, whether it is a brand name or a specific type of product or item that a company offers. The reason this is important is because sometimes developing keyword research can involve getting more specific, in order to uncover keywords for products where there isn't as much competition.

Search Volume and Competition

Another concept to learn about is search volume. When you're thinking about search volume, you're simply saying, "how much demand is there for these keywords?". This is a good question to ask, because even if you think of keywords that relate well to your products, it doesn't necessarily mean that there's *demand* for the keywords. In other words, just because it's a good keyword, doesn't mean that people are actually searching for it as much as you'd like. Part of the process of keyword research is checking to see how much search volume there is for particular keywords, which we'll look at.

The other thing to remember, as you learn about keywords, is that there's competition. This applies whether you are writing ads (using Google AdWords—see *Introduction to Search Engine Marketing and AdWords*) or you're doing search engine optimization.

When you are running ads, the competition is a bidding environment, where you decide how much you are willing to pay for a click on an ad. There is competition around particular keywords. For example, if two drone manufacturers want an ad to appear about their company on Google when you type in "racing drone," there will be competition for that keyword phrase.

There's a similar concept at play for SEO, and there is competition for keywords there as well. If you think about search results and the web sites that come up, you can see how everyone would probably like to appear on the first page of search results. That's the best place to be.

For example, at the time of this writing, if you type in "sunflower seeds", you'll see in the fine print at the top that there are about 6.5 million search results on Google, from web sites around the world!

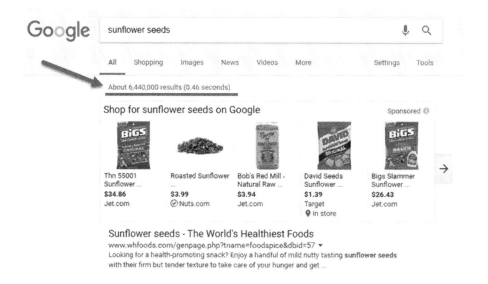

And at the bottom of the first page, you can navigate to more pages of results.

In theory, you could page through 6.5 million pages of search results, but you're probably not going to do that, and the point is that the higher and *earlier* you appear in the search results, the better. In general, the initial goal is to appear somewhere on the first page, or within the first 100 results, and then to work on increasing your rank from there.

SEO is the process of maintaining your rank and ideally increasing it. It is fair to say that aside from thinking about technical things like using meta descriptions (see Chapter 9) and doing keyword research, that developing content and adding it, to strengthen and reinforce keywords, makes sense. Even if you didn't get super specific about how many keywords were in a particular article you added, for example, you might still benefit from being aware of keywords, and either writing material or working with writers at your company (or freelancers) to develop written material that is related to keywords. For example, if you are a drone manufacturer that offers drone kits, you could write an article on "The Top 10 Considerations for Drone Kits" and write it naturally, where the phrase and keywords appear in the article.

The easiest way to think about gaining experience with keywords is to think about it as a mindset—as you are working on content or working with people who develop content, it can become a conversation to think about keywords and how they will boost marketing objectives. There may already naturally be content that will be added to a site, but an SEO person will instigate that through blog posts, articles, etc.

Doing Basic Keyword Research

We're going to look at two options for basic keyword research, based on a scenario of a sunflower farmer who wants to sell some sunflower seeds. Farmer Sally sits down at her computer and has a lot of options, but she likes to use a few different tools. One easy one to use is Wordtracker (https://app.wordtracker.com/).

You get five free searches to start and you get a risk-free seven-day trial of each access plan.

To give it a shot, try typing in the phase "sunflower seeds" and see what happens. It will look something like this.

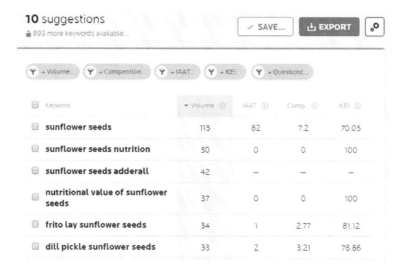

Click on any of the individual "i" icons in the column heads to see more information.

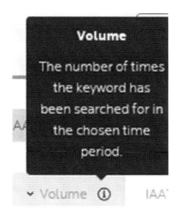

In general, what you see is a guess about traffic, in thousands, in the volume column, and an estimate of the amount of competition. This tool also provides suggestions of other keywords to consider (sunflower seeds nutrition). These suggestions are called "long-tail keywords," and they are basically a little bit more specific. That means they could potentially pick up more traffic.

Wordtracker in general is one of those services/sites/tools that is worth becoming familiar with as you do keyword research and learn about the process. Aside from trying the tool to identify keyword opportunities, Wordtracker also has some good learning materials:

`https://www.wordtracker.com/academy`

The screenshot only shows two items, but there are several areas that are worth looking into:

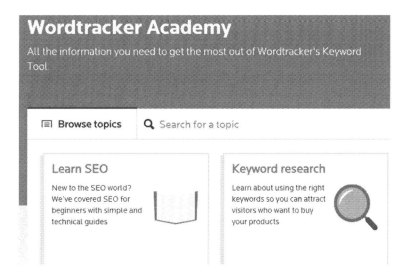

The Learn SEO category has some good lessons:

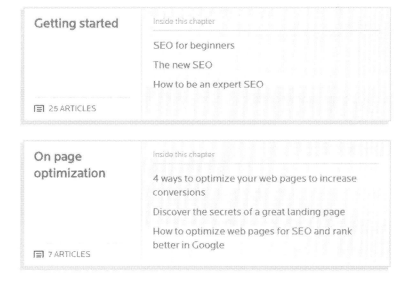

AdWords Keyword Planner

Google AdWords keyword planner is a common tool for keyword research and it highlights the connection between SEM (AdWords—search engine marketing) and SEO. In order to use it for SEO, you need to start an AdWords account. The easiest way to do that is to have a Gmail address, but if you like, you can use a non-gmail address to establish a Google account.

To start an account, go to www.google.com and start the process.

At some point in the process, in order to "initiate" the account, you will need to enter billing information. I say go ahead and try making an ad campaign using the wizard that they offer as part of account startup, just for the experience. (Maybe consider my book, *Introduction to Search Engine Marketing and AdWords* for more information.) AdWords is outside of the scope of this book, so if you don't want to try running an ad, make sure that you click on the Campaigns link at the top of the screen:

Then click on the green button to pause your campaign:

It should look like this:

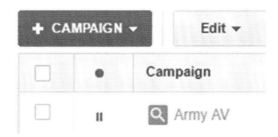

By jumping through those hoops, you'll have an active AdWords account, and you'll be able to use the integrated Keyword Planner tool.

Trying the AdWords Keyword Planner

To try the AdWords Keyword Planner, log in to AdWords and select Keyword Planner from the Tools menu:

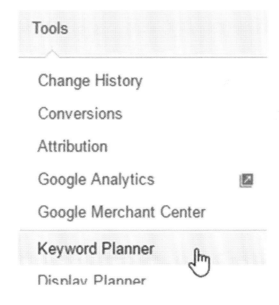

Before you start using it, I recommend exploring some of the articles that appear on the screen:

Before you begin

How to use Keyword Planner

How to see your organic data

Building a Display campaign? Try Display Planner

To try it, click on "Get search volume data and trends":

Keyword Planner
Where would you like to start?

Find new keywords and get search volume data

▸ Search for new keywords using a phrase, website or category

▸ Get search volume data and trends

▸ Multiply keyword lists to get new keywords

The search volume window has a variety of options. In general, I suggest going with the defaults as you learn the program.

Try entering a keyword:

Then click the Get Search Volume button:

You should see something like this:

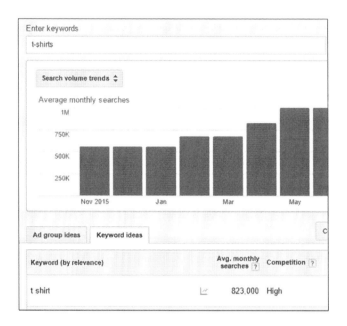

There's a lot of interesting information, but the main purpose is to get a general idea of how much search volume there is. In the earlier search window, the "targeting" was to "all locations," meaning the entire world, so narrowing that down when you do a search is a good idea unless your web site is targeting the entire world.

You can also see the relative amount of competition, which is important to be aware of for SEM and SEO purposes.

At the top, you can enter new words or modify an existing search (to change the targeting):

With that, you have entered the sophisticated world of keyword research using a core tool, the AdWords Keyword Planner! Congratulations!

Now for a bit of fun.

Activity: Research Keywords

Pretend you own a T-shirt business and are looking to generate ads on Google and work on search engine optimization.

Think of five keywords to start with and use AdWords Keyword Planner to research them. Then, either think of five more keywords or use Wordtracker. Type one of the first five keywords and see what Wordtracker suggests. Register for the free trial if you have to.

You can also just start typing in keywords in Google and see what "auto-suggest" comes up with:

Make a Keyword Target List

As part of the activity, make a keyword target list that includes search volume and amount of competition.

Make a Keyword/Topic List

Take a look at your list of keywords and then generate ideas for titles and topics of blog posts or articles you could write that relate to each keyword. Think about how you could place a keyword into the title of the article, as well as into the text of the article at least once.

This exercise represents the intersection of SEO and content development, where you're going beyond just developing content and you're thinking about how keywords and content relate. In this case, you're starting with keywords and developing keyword-driven content.

To take this exercise one step further, try reviewing the articles from the last chapter about some of the steps you can take to do an SEO audit. Try some of the related tools and pretend that you were helping a new t-shirt business that has a web site but hasn't done SEO before. Make a list of the things they need to set in motion. Check out these articles:

https://moz.com/blog/how-to-perform-the-worlds-greatest-seo-audit or http://tinyurl.com/mozaudit

and

http://www.plusyourbusiness.com/first-9-seo-tasks/ or http://tinyurl.com/9seotasks

Don't worry about doing all the steps; just make a rough plan of the kinds of things you need to do, including the tools you plan to use.

The list probably would/should include using webmaster tools, which we look at in Chapter 7, so don't worry about that too much. Just take a look at things at a high level for now.

Have fun!

Additional Resources

Here are a few helpful resources that are worth looking into as you explore keyword research:

- Beginners Guide to SEO: http://moz.com/beginners-guide-to-seo/keyword-research or http://tinyurl.com/mozbegi
- Keyword Volume Tools: http://moz.com/blog/keyword-volume-tools or http://tinyurl.com/mozkeytools

Check out these tools as well. It's worth it to try their free trials at least, so that you can say you've used them:

- www.keywordspy.com
- www.seomoz.com

Finally, you can find free SEO tools at https://moz.com/tools.

Conclusion

Congratulations on exploring the "10% of SEO" outside of content! In the world of technical tweaks, the larger the company is, the more revenue is at stake, and the larger the effort that companies generally make with SEO. There's no one rule about how to do it or what kind of resources to allocate, but it's fair to say that as a top skill to get people hired, a full-time SEO person might spend most of their time doing the technical tweaks and measuring performance. The technical side is definitely important, but it's easy to get carried away and lose sight of the equally or more important fundamental need for good content.

If you find yourself apply for a full-time SEO position, you'll want to see if it involves content development. If it doesn't, you'll probably want to ask them why in an interview and see if the digital marketing team uses copywriters to develop content to boost SEO results. If they don't, you might suggest that they consider it at some point.

In a "multiple hats" role or if you are doing SEO yourself, find some way to make sure you are prioritizing and developing content alongside the technical things you do. The technical tweaks are only part of the picture.

Try SEO

In order to get some practice with SEO principles, this chapter uses a few simple, live examples, starting with Blogger and moving on to Google Sites. If you haven't already, read Chapter 2 first. You'll learn more from this chapter if you create a blog at blogger.com and a super simple web site at http://sites.google.com.

For the content of your blog post, I recommend writing a simple blog post that contains 2-3 paragraphs and an image, on any topic you've learned about so far. You can then use that blog post as an example for these activities.

If you want, you can copy the information from my example blog at http://toddsmarketingblog.blogspot.com.

Simple SEO: Blog Post

As you learn more about SEO, you'll start to think of things you can do when you work on a site for the first time, such as:

- Determine if it is configured for search
- Adjust the meta description for the site
- Adjust the meta description for individual pages/posts

Additional things you can try include:

- Integrate keywords into blog post title
- Integrate keywords into blog post text
- Integrate keyword(s) into alt text for images

© Todd Kelsey 2017
T. Kelsey, *Introduction to Search Engine Optimization*, DOI 10.1007/978-1-4842-2851-7_5

These latter options include going beyond "natural" content writing and thinking about how you can "insert" keywords. There's nothing necessarily wrong with that approach, just don't go too far. In many cases, you might not even need to change the title of a page or an article, because if it is named after the topic it covers, chances are it has a relevant keyword in the title already.

Note You're welcome to access and copy the text from my blog posts (`http://toddsmarketingblog.blogspot.com/`) for practice if you like.

Next we look at an example of how to enter the overall meta description into Blogger.

Blogger Overall Meta Description

Log into Blogger, scroll down in settings, and choose Search Preferences.

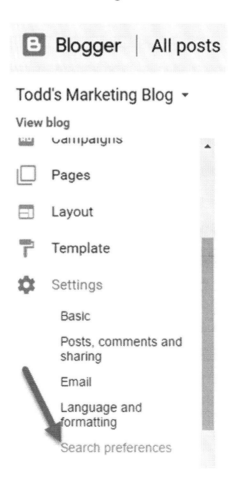

Over on the right (and anywhere else you see question marks in any Google product), click the question mark.

There's often helpful information, and sometimes links to more:

Configuring the meta description in Blogger is usually a one-time thing. Click Edit.

Click Yes.

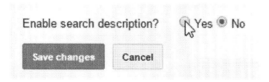

Then type a description. Notice the 150 character limit.

Something like this:

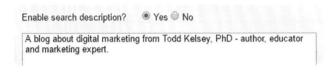

For grins, you might use this tool (and file it away) to determine how many characters you are typing in:

https://charcounter.com/en/

Then click Save Changes:

In theory, the description you chose will eventually appear in the Google search results.

Post-Level Description

What we just did was for the overall site, but what about individual posts or pages? That's worth looking at too. In that case, we can go into Blogger and edit an individual post.

On the right, in Post Settings, click Search Description:

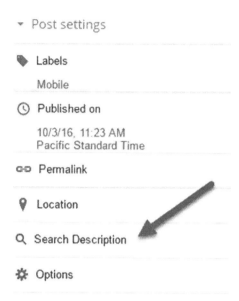

Enter something that describes the post and click Done:

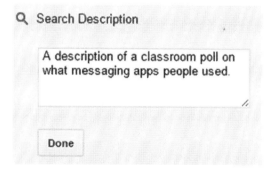

At the top, don't forget to click Update:

Site-Level SEO: Google Sites

Depending on what platform you use, adjusting HTML tags for SEO may be very easy, or it might require some technical assistance or research. In this chapter, we're looking at very simple examples, and there are limitations to the platforms, but you can get the basic idea.

Google Sites is not that different from Blogger, but it's designed more around creating a conventional web site with informational sections, as opposed to a diary or journal format (i.e., a blog.) The difference blurs, but it's good to see how things might be different using a different system, and as you'll see, you should always be prepared for the unexpected. (Hold on to your hats!)

We're going to adjust the title tags on Google Sites (by setting page titles). If you haven't yet create a Google site, please do so, at `http://sites.google.com`.

■ **Note**   In this chapter we use the New Sites option. Depending on when you read this, you may have an option for Classic or New Sites, or only be able to choose the new approach.

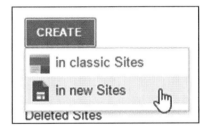

You'll end up with an interface that looks something like this:

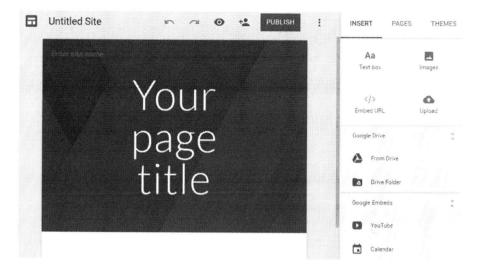

Click on Untitled Site and replace it with something like "Pet Store":

Click Publish:

This is where you choose the web address. You don't need to do this every time. You'll have to experiment until you find something that isn't taken.

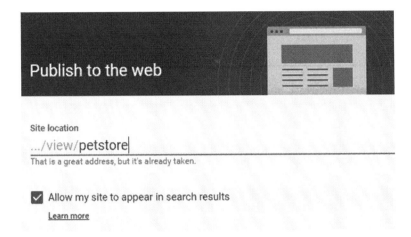

You can include numbers, use your name and so on, and when you find something available, a little checkmark will appear.

Next, click the Publish button and then the View link that appears at the bottom:

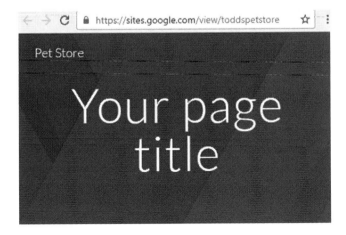

Your quickie-doo site should look something like this:

Next, on the right (back in the editor, maybe under the original browser tab), click on Pages:

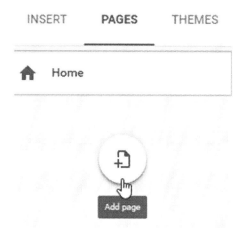

You should see something like this, where you can add pages.

Try adding some new pages. Your chosen title should be reflected in the <title> HTML tag. If your planning process included SEO when you were creating a new page or new web site, you may have thought about what to name the pages, such as Dogs or Dogs for Sale. (People are more likely to type in "dogs for sale" than just "dogs" if they are looking to buy a dog).

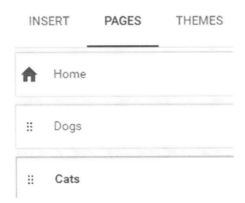

Activity: Site Title Optimization

In this section, we're going to try something, but it might not work. With some systems you might not be able to update the <title> HTML tag when you rename a page. Say you created a page called "Dogs" but you want to update it to "Dogs for Sale". You might be able to rename the page without having to edit the HTML. It's worth trying.

Scenario: Imagine you are providing a specific product or service and thinking about how the titles of the pages about the products or services can include keywords. For example, try t-shirts or a pet store.

Try making 1-3 pages based on your chosen product/service/keyword. For example, create a home page, a page about cats, and a page about dogs.

Now, when you try publishing it, view the site.

Select the desired page:

Then view the code. In Windows, right-click and on the Mac, Ctrl+click. Then choose View Source (or View Page Source):

Then press Ctrl+F to activate the "Find Text on Page" function. Scroll down until you see something like this:

Oh no! It's HTML code!

Don't be alarmed!

Take a closer look for something like this:

`<title itemprop="name">Pet Store - Dogs</title>`

What you see is that the title of the page is reflected in the HTML code. Sometimes it makes sense to emphasize keywords in the page titles. Sometimes it makes sense to learn about the 200 other or so SEO ranking factors.

Oh no! 200 ranking factors!

The good news is that you don't need to learn all of these factors. Remember how content is 90% of the battle? Don't worry about these keywords so much. Instead, start looking at industry resources like www.searchenginenews.com and www.moz.com to keep an eye out for important changes to the way Google crawls for keywords.

Because sometimes things change.

Okay, back to the dogs.

I could use my web platform to change the title of a page:

And then publish and view the page:

I might even expect that when I view the source code:

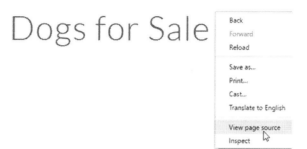

The `<title>` HTML tag will be updated, right?

Well, let's take a look and see.

For you, it might work. At the time of this writing, Google Sites, both New and Classic, do not appear to update the `<title>` tags when you change the page title. Doh! Google may update this feature by the time you read this, or they may have a great reason for not including that feature. (Although I can't think of one).

The moral of the story is, you've been getting a free ride in this book, in terms of the simplicity of the platforms, but at some point you might need to dig a bit into technical things, and I recommend learning more about HTML. (See `www.casamarketing.org`, look for content and SEO, and then look for SEO and HTML).

Activity: Learn HTML the Easy Way

Here's a little bonus—a secret way to learn some HTML.

If you go back into Blogger and open a post, you can do something like type in a word, like the word "bold," for example.

If you published the post, Blogger would use HTML code to make your page look something like this:

If you are hardcore, you can use your Jedi powers to look at the HTML and find an occurrence of "bold" like this one, with no tags around it.

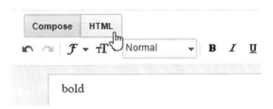

To view the HTML, you can edit your post and click on the HTML tab:

Okay, I don't see any code here, but that's because it's just plain text.

Back into compose:

Now select the text and click B to make it bold:

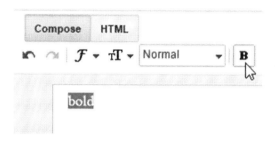

Click on the HTML tab again:

What's that? I see some little tags. It's the <b> tag, which codes for bold. There's also an ending tag, </b>, which tells the browser to stop applying bold to the word or phrase.

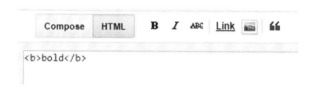

So what you can do, using this "secret" way to learn HTML, is go back and forth between Compose and HTML and try different things, like formatting the text, adding a link, making the text centered, larger, smaller, a different color. Try inserting images other kinds of things. Each time you make a small change, switch over to the HTML tab to see what's going on. It's probably more fun than reading the articles I suggested earlier. Go ahead and do this fun part first, and experience the power of the secret way to learn HTML.

For Further Review

It's important in this industry to always keep researching and keep up with the times!

- Other than the SEO and HTML article on www. casamarketing.org, here are some other things to take a look at, consider, and perhaps even blog about. (It's okay to link to other resources in a blog post!)

- Review number 6 in this article: http://www.plusyour business.com/first-9-seo-tasks/ or http:// tinyurl.com/9seotasks

- Read this title tags article: http://searchenginewatch. com/sew/how-to/2340747/title-tags-seo-3- golden-rules or http://tinyurl.com/seogolden

Conclusion

Okay, you deserve a break now. Congratulations on wading through some basic SEO optimization, as well as getting your feet wet with HTML. Don't be intimidated—just start small and learn a little bit here and there. Before you know it, your confidence and skills will grow.

Analyze How Things Are Going

This chapter looks at some of the ways you can measure the performance of your SEO efforts. It's good to do, whether you are working on your own site, to justify the time and effort spent, as well as to measure the ROI if you are paying someone, or if they are paying you!

Ultimately the ideal spot to reach is the first page of search results for a given keyword, and the top spot. Often SEO "rank" programs will only measure "down" to the 100th search result. At the beginning, if you're a new or small web site, it may be some time before you can crack the top 100 search results for a keyword, but it is still worth trying, and regardless of the exact rank, it's absolutely true that adding relevant, quality content to your site will make it stronger no matter what, regardless of the rank on Google, for the people who do visit it.

There are different philosophies about what kind of effort to put into SEO versus SEM, and there's no one correct answer for everyone. SEO is likely to be an important skill and a good source of income for quite some time, and it stands to reason that if you focus your SEO efforts around developing quality

© Todd Kelsey 2017

T. Kelsey, *Introduction to Search Engine Optimization*, DOI 10.1007/978-1-4842-2851-7_6

content, it's unquestionably a good investment, even if you don't get the exact rank you want. It's worth saying that there are those who recommend making sure your SEM efforts are off and running first, where you can jump to the top of page 1 with the right bid, for example. But even SEM efforts are affected by quality content—when you deploy content for SEO, it can also boost SEM efforts, through a metric called your quality score. (See my book, *Introduction to Search Engine Marketing and AdWords,* for more information.

But in the end, is SEO worth it? Absolutely. Make sure to learn how to measure the performance.

Fun with SEO: Check Your SEO Rank!

One way to measure and monitor the performance of SEO efforts is to look at the rank of your web site for particular keywords. There are a number of tools available—some free, some free trial and then pay, and so on. Here's one to try:

SEO Centro: `http://www.seocentro.com/tools/search-engines/keyword-position.html` or `http://tinyurl.com/seocentrokp`.

You can enter any web site, but if you like you can follow along in the example:

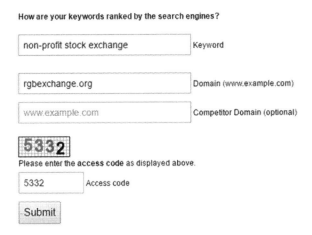

You enter a keyword and type in the web site's name. It checks what is going on.

Main Result

	Positioning History
Report ID:	#8433738
Load Time:	1 second(s)
Report Generated:	2016-12-18 01:58:39 GMT
Domain to verify:	rgbexchange.org
Keyword to verify:	non-profit stock exchange
Related Keywords:	n/a
Spell Check:	ok

rgbexchange.org

Search Engine	Placed	Rank	First Url Found
Google	yes	18	http://www.rgbexchange.org/
Bing	no	n/a	n/a
Yahoo!	no	n/a	n/a

Google Result

Rank	Url	Keyword(s) found in Title	Domain	Url
1	http://npoex.strikingly.com/	no	no	no
2	http://smallbusiness.chron.com/can-non-profit-corporations-i...	33%	no	67%
3	http://effective-altruism.com/ea/u3/nonprofit_stock_exchange...	100%	no	67%
4	https://www.facebook.com/npoex/	67%	no	no

If a web site ranks in the top 100 for a given keyword, it will show the rank. In that case, your strategy might be one of the following:

- *Maintain the rank:* There is competition for online keywords, and it will continue to increase as the Internet grows and the world becomes smaller. Maintaining existing rank is a good goal.

- *Increase the rank:* This might involve a strategy of doing some of the technical tweaks to make sure all the appropriate basics are done, and then work on developing and expanding keyword targets, in partnership with any search engine marketing efforts. You try to increase the rank through consistent deployment of quality content, over time. Not just content, mind you, but quality content, based on finding out and even testing what your customers *actually* are interested in.

Here is an article with some relatively recent free/free-trial SEO checkers, which is worth checking out. It's called Shout Me Loud: https://www.shoutmeloud.com/5-excellent-websites-to-check-keyword-ranking-in-google.html or http://tinyurl.com/sml-seo.

Among other tools it talks about SEM Rush, which is a pretty standard SEO tool that ties into SEM. There is a free 14-day trial and it's probably a good investment. Compare the free trial/paid tools to any "entirely free" tool, and you will get what you pay for.

Fun with SEO: Hunt for Duplicate Content!

Duplicate content is a large no no. You never want to have multiple pages on your site with exactly the same content, unless it's unavoidable (as it might be on some larger e-commerce sites). The strength of content is in its uniqueness. It is an important thing to watch for, as is fairly common (pages copied, forgotten, material paraphrased, pages migrated from one place to another, redundancy—you name it! It happens). It's common enough to be a part of a good SEO audit.

Hunting for duplicate content can be fun and rewarding and it's also "low hanging fruit" as part of an SEO audit. It also gives you more time to increase your SEO rank.

Siteliner.com is a good tool for scanning for duplicate content.

Explore your site.

http://		Go

Find duplicate content, broken links, and more...

Give it a shot. Pick your nearest university and see what happens.

www.ben.edu

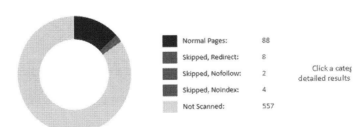

MAIN MENU
Summary ▸
Your Pages (88)
Duplicate Content (12%)
Broken Links
Skipped Pages (14)
Related Domains (10)

We're scanning your site...

Scanning your site... 102 pages scanned of 659 found.

college-of-science/index.cfm College of Science | Benedictine | Chicago | Catholic Universities

Your Pages – Click below to see detailed results for your pages:

Normal Pages:	88
Skipped, Redirect:	8
Skipped, Nofollow:	2
Skipped, Noindex:	4
Not Scanned:	557

Click a categ
detailed results

It will look for duplicate content on your site and give you a way to look back and see if it is redundant or necessary. The reason duplicate content is a big deal is because Google can punish you for having it if it thinks you're using it as a way to get higher rankings nefariously. Also, good content is worth it, in it's own right.

Reporting can include potential redundant content and part of the SEO process would be taking content away. Wacky, right? Here you are wanting to add content and there you are taking it away. But not all content is created equal, especially if it is duplicated or "cloned," as it were.

Some people like clones I, suppose. But not Google, at least when it comes to content clones.

Other Analysis: Analytics

Okay, in the end, "rank" is one of the main things to monitor, but not the only one. You saw how an SEO audit can reveal duplicate content; that's another thing to monitor, periodically even, especially if you are at a larger company. Eventually, sooner or later, you will want to get into analytics.

I understand if you have any feeling of intimidation about analytics. Case in point, once upon a time, the author of this book was in a band, and that band was on tour and rode around the country in a fancy bus.

But what happened to that bus? Well, we didn't have a hit radio single, and I landed in a cubicle at an Internet startup company, like a fish out of water. If you'd asked me whether I had an interest in analytics, or Excel, or numbers of any kind, or even marketing, I would have laughed out loud. (What, me? I'm a creative person.) But being a creative person doesn't always pay the bills. I continue to at least try to be creative, and that's one of the reasons I like inserting silly pictures into the books I write.

Over time, as I experienced the phenomena known as being *laid off*, I realized that there are parts of companies and things they do that *make* money, and then there are things that *cost* money. Marketing, if it's done well, is one of those things that helps make money, and I gravitated toward it. I also realized that even creative people can gradually get more comfy with things like Excel, and numbers, and analytics. In other words, learning numbers and analytics is a good idea even if you're not an accountant.

Analytics can give you superpowers!

(Image designed by Freepik.com.)

With these superpowers, you can:

- Help a business not waste money
- Help a business decide what to invest in
- Help a business make money
- Keep your job
- Keep a client happy
- Expand your career
- Have interesting things to say in meetings
- Sound like you know what you are talking about
- Know what you are talking about

And many more.

So, who wouldn't want to do that, right? If you remember, analytics is a core area in digital marketing.

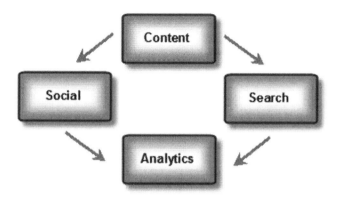

It's worth taking time to learn these things outside of SEO. There are many free resources out there, including a Google Analytics Certification and free study material from Google. (And there's my *Introduction to Google Analytics* book too).

For SEO, it's good to note that Google Analytics can provide a foundation for giving insights about some parts of SEO. When it is set up correctly and connected to Webmaster Tools (now called Search Console), it can provide you with fairly detailed information right from your own site that can monitor the way individual keywords perform.

Before you try it, though, we will discuss the Holy Grail.

The original Holy Grail was a cup made out of Gold. But the next best thing to the original grail is to be able to measure how much gold you get back, when you invest your gold. Another name for tracking your gold is conversion tracking.

Fun with Conversion Tracking in SEO

This book is supposed to be an introduction, but this whole conversion about tracking holy grail thing is important enough to at least mention. It's so important that it could:

- Get you a job

- Help you keep your job

- Help other people keep their jobs

- Help everyone get a bonus

- Help people get a holiday bonus

- Make all your dreams come true

Conversion tracking essentially allows you to answer a manager or client when they say, "Okay, so we spent a lot of money. How much money did we make?"

Let's say you're at a company called Round Table that offers specialized products for Arthurian knights.

King Arthur comes along for the Monday marketing meeting and asks about the ROI.

King Arthur: "Okay, how did we do with radio advertising this week?"

Lancelot: "Well, we had a lot of impressions."

King Arthur: "Okay, but how many people purchased our knightly protection service based on radio ads?"

Lancelot:"Um . . . ."

Let's consider another knight, named Gawain.

King Arthur:"Okay, how did we do with pay per click advertising on Ye Google?"

Gawain:"King Arthur, because of our goodly conversion tracking, we were able to ascertain that two townsfolk of good repute did indeed buy our knightly protection services."

King Arthur:"And how much did we pay Ye Google?"

Gawain:"King Arthur, we did pay Ye Google a fee of 10 gold pieces for the magic clicks on our advertisements in the town, and we generated 100 gold pieces in return."

King Arthur:"I like this conversion tracking. Why aren't all the knights of the realm doing this?"

Lancelot, grumpily:"King Arthur, it takes effort. I'd rather be off riding and rescuing than doing configuration on conversion tracking."

King Arthur:"Ah well, the royal treasury is not what it used to be, what with Saxon rampages and other unexpected events. Going forward, all knights shall use conversion tracking."

Knights: ::groaning and grumbling::

King Arthur:"Yes yes, but I've decided that it will be our new Holy Grail!"

Knights: ::applause::

So the moral of the story is that conversion tracking is important.

All this to say that it does require some effort, whether with SEM or SEO. We're not going to dive into that in this introductory book, but it's important enough to mention, emphasize, and consider.

With SEM, in a nutshell, you can set up conversion tracking, and if you're selling something it typically means putting a piece of code on an order confirmation, and it beams a little signal to Google saying that a sale has been made. When all the pieces are put together, you can actually track how much revenue is generated from an individual advertisement or keyword. This is a tremendous thing, and it's one of the reasons Google is one of the most profitable companies on the planet—because they figured out how to do trackable advertising.

With SEO, it's similar. With some effort, with accounts linked and code in the right place, using perhaps SEM/AdWords, you can measure your ranking on SEO. You can also:

```
measure how much money is made by organic
keywords
```

Yeah, sorry, that's quite a mouthful. But it's a good thing to chew on.

What it means is that with conversion tracking set up, you will be able to see how much revenue resulted from:

- People searching for something on Google with a specific keyword

- People seeing your organic search result (as a result of your SEO effort)

- People buying a product based on that keyword

To repeat, you will be able to see how much money resulted from SEO efforts. Exactly. It's a really good, strong thing to do. Not only can it help win more attention, more serious attention, or attention at all from a client, manager, etc.—but it also results in more investment. It results in more scrutiny, but it gives you a strong leg to stand on.

In other words, when clients or managers don't really understand SEO, and do it on faith because someone said they should, or because they read an article about it, they may be willing to invest in it, but still don't take it seriously. If the budget is reduced for any reason, and the heat is on, then that line item expense (or job) could be on the line. How much better, then, to flip the tables, measure things out, and work toward proactively gaining scrutiny, by demonstrating how organic SEO can actually *make money*.

It's also a very real possibility that consistent, longstanding efforts at SEO could generate revenue and profit. Even if it doesn't start out that way, even if it is a long-term investment to increase rank or customer appreciation of the quality content on a site—even if it's a while before you can actually prove a sale is coming through organic keywords—it is absolutely worth exploring.

Conversion tracking can make that happen. Google "conversion tracking adwords" or "conversion tracking of SEO in Google Analytics" and things like that. It requires some hoops, it's outside of the scope of this book, but I've had my say, and I'll leave it at that.

Just remember that it is the Holy Grail. Consider blogging about the Holy Grail.

A recruiter or potential employer or potential client might actually notice.

Set Up Google Analytics: Blogger, Google Sites

So let us return to simpler things in this introductory book. Google Analytics is very easy to set up in Blogger and Google Sites, so I invite you to do so, in order to get more familiar with it. We'll look at one of the basic SEO-related reports you can run without having to set up conversion tracking.

Setting Up a Google Analytics Account

Follow these steps to set up your Google analytics account.

1. Go to http://google.com/analytics and sign in and/ or set up an account.

   Access Google Analytics

   If it's the first time you're using it, Google may guide you through a wizard to set up a web site. Enter your blog address for the web site, such as http:// toddsmarketingblog.blogspot.com.

2. There's more than one way to add a web site to Google Analytics. If you've used it before, you can access the Admin menu:

   Admin

Then select the Account drop-down menu:

And click Create New Account:

3. Next, you enter your web site information.

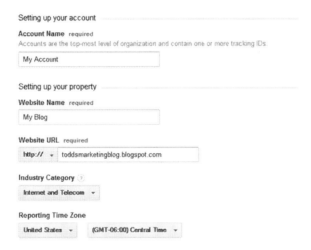

4. Give your site a name, enter the address/URL, choose a category, and set the time zone. At the bottom, click Get Tracking ID:

■ **Note**   The Tracking ID is the piece of information that you put on your web site, in order to connect it to Google.

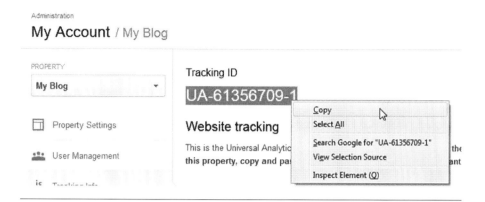

5. Copy the tracking ID into memory. It will look something like UA-61356709-1. (It will be a specific, unique code for your site/blog.)

6. Next, log in to your Blogger blog and go to Settings:

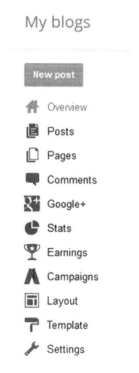

Then select Other:

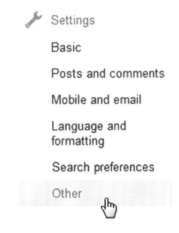

7. Enter your Google Analytics ID (Note: it will be a specific number you got from your own account, not the one shown in this image).

8. Click Save Settings.

Now you're all connected! Potentially immediately, or certainly after a few days, you'll be able to go into Google Analytics and look at detailed information about your site.

Create a New Property for Each New Web Site

▓ **Note** You can have more than one Google Analytics account, such as one for each web site, but generally people use one Google Analytics account and include all their web sites in that account.

To track a new web site, you can add a new "property" to an existing account. For example, go to Google Analytics ➤ Admin.

Click the Property drop-down menu to create a new property. Think of this as digital real estate.

Then you can enter the information for each web site/blog you create and want to track. If you are working with a client, you will probably want to create a separate account. If you are working on your own web sites, then you probably will want to just add new properties for each site/blog you work on.

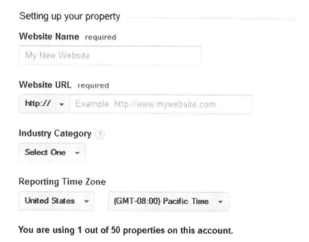

▓ **Note** For each new site/blog you want to track, you need a tracking ID—that's the piece of information that you use to connect your web site to Google Analytics.

Google Sites: Enter the Google Analytics Tracking ID

▓ **Note** This step assumes that you've created a new property in your Google Analytics account, and that you have a new tracking ID. You need a separate, unique tracking ID for each site/blog. You can't use the same one in multiple sites/blogs.

At the time of this writing Google is transitioning from Classic Sites to New Sites. At the moment only the Classic Sites allows you to connect to Google Analytics. By the time you read this the New Sites might allow it too. Just give it a shot, and if it doesn't work, try getting in touch with your Inner digital analytics detective, and try doing the equivalent in a free Weebly site. You make a free Weebly site and then research how to connect it to Google Analytics.

For Google Analytics with Classic Google Sites, go to `https://sites.google.com/` and select your site. Go into Settings (the gear icon) and select Manage Site:

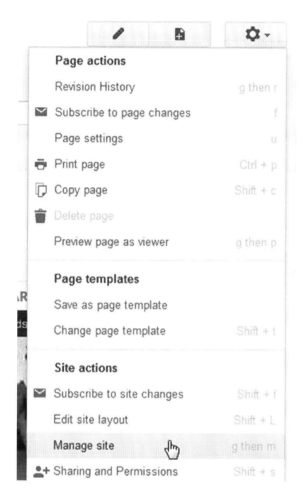

Then scroll down, make sure the Statistics drop-down menu is set to Use Google Analytics, and enter your property ID/tracking ID. (Note: it will be a specific number you got from your own Google Analytics account, not the one shown in the following image).

Click Save.

Now you're all connected! Potentially immediately, or certainly after a few days, you'll be able to go into Google Analytics and look at detailed information about your site.

Fun with Analytics: Using Your Superpowers

Now that you have Google Analytics and it's been running for a few days (wait a few days, and also try to promote your site by sharing it on social media, if nothing else, to get some traffic), it's time to take a look. Follow these steps:

1. First, log in to Google Analytics.

2. Check the Date Range to be aware of what time frame you are looking at it, and change it if desired:

3. Click on All Web Site Data:

4. In order to find the right report, look for the search box:

5. Try typing in All Traffic Channels and select it from the pop-up menu:

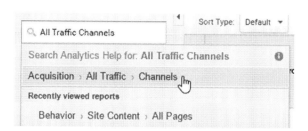

You will get more out of it the longer you have Google Analytics installed, and the longer you've been promoting your site (so that not only are people visiting it based on your initial promotion, but ideally searching for it on Google). If you want to see what it's like, but don't have a web site, e-mail me at tekelsey@gmail.com and I will invite you to the RGB Foundations Google Analytics account so you can take a look around. (No promises mind you, but it's an option.)

Here's the kind of thing you'll see. Even if you're not doing conversion tracking, and just want to see what proportion of traffic is coming from organic searches, it's cool:

☐	1.	Organic Search	**6,149** (87.58%)
☐	2.	Direct	**712** (10.14%)
☐	3.	Referral	**118** (1.68%)
☐	4.	Social	**42** (0.60%)

Another thing to consider with SEO is to set the goal of increasing traffic from organic search over time, based on the amount and the proportion of traffic. In some ways, SEO is like an investment that becomes a source of ongoing income—you do have to maintain it, but when the right things are done and everything falls into place, it is possible to generate traffic, sales, etc., without having to spend money. Managers and clients will like this. (Technically, you are spending money by having someone work on SEO, but it at least sounds good. And there's some legitimacy to it.)

Getting Google Analytics up and running is a good thing, and here's an article for further reading (and blogging about) if you're interested:

> *Analysis techniques for Google organic search and SEO:*
> https://support.google.com/analytics/
> answer/3306157?hl=en or http://tinyurl.com/
> seo-googa

You might find reading a book helpful, such as my book, *Introduction to Google Analytics* (Apress, 2017).

Conclusion

Congratulations on launching into the deep world of measuring performance with SEO. It can be a journey, and you might have to do a bit of exploration, especially for the Holy Grail (conversion tracking).

But there can be a lot of treasure involved, and I highly recommend it.

Don't be alarmed by analytics. Just think of it as treasure hunting. Even if you're not a numbers person, it's rewarding. If I can make the journey, you can too. Remember, I'm just some rock 'n' roll guy.

From one superhero to another, best wishes with your new found superpowers!

(If you want to procrastinate before you read the next chapter, feel free to check out these pics of my past life in rock 'n' roll. See `http://tinyurl.com/tkrocknroll`.)

Of course, feel free to reach out to me on LinkedIn: `http://linkedin.com/in/tekelsey`.

Explore Indexing and Webmaster Tools/Search Console

This is a short chapter, but an important one.

Before you do anything, let's take a moment to return to the "How Search Works" video by Matt Cutts. If you haven't seen it yet, you should, and even if you've seen it before, it will be a helpful refresher for this particular chapter.

Check it out again at https://www.youtube.com/watch?v=BNHR6IQJGZs or http://tinyurl.com/howsearchworks.

One issue to pay attention to and consider is how Google crawls and *indexes* the Internet. How does Google know your web site is there? There are any number of ways that it will find your site, based on what platform you're on, if you post on social media, etc. In some cases, it's good to verify that everything is okay. It would be nice if everything on the Internet worked seamlessly, but

© Todd Kelsey 2017
T. Kelsey, *Introduction to Search Engine Optimization*, DOI 10.1007/978-1-4842-2851-7_7

that's just not the case. Because of all the different systems out there, and because of technical and human errors, even web sites that have top technical staff and a lot of resources sometimes end up in a situation in which the site doesn't speak to Google as well as it could. It's good to monitor this, and it's a basic thing to do as part of an SEO audit.

We'll start with a really simple way to check things that doesn't require any special tools.

Fun With SEO: Check Indexing with Google

One simple way to check out how Google sees your web site is to go to Google and type this in:

```
site:yourdomain.com
```

Google lists all the pages it sees. Especially with larger sites, it's good to get an idea of approximately how many pages there are, and then compare this number to what Google is seeing. If there's a difference, you have a problem. You want Google to be able to see everything that is meant to be public.

For example, you might see something like this:

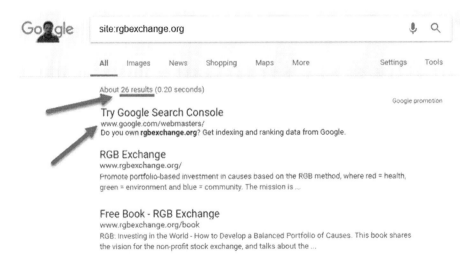

What that tells me is that Google sees 26 pages. If you scroll down and click through the pages, you can see each one. In some cases, you might see pages that you forgot about or that are not supposed to be public!

The indexing example is that simple. It can be, should be, part of your SEO audit. It's one of the simpler things you can do if you are doing SEO freelance and want to offer a free SEO audit to potential clients.

Okay, hold on to your hats—adventure awaits! Get ready to explore new worlds!

Get ready to enter the land of *webmasters*!

Explore Google Webmaster Tools/Search Console

Back when it was harder to create and publish web sites, you had no choice but to get technical. Although some people did everything, you often had a distinct role of webmaster, which was part Jedi Master, part doctor.

In some cases, there still is a person who works with the web site as a full-time job, more often than not, on the very technical side, such as maintaining the web server and related systems. The larger the company, the more complex the web site. There might be a whole team that works behind the scenes on the very technical end of maintaining web sites.

Many companies, especially smaller ones, use a "managed service" such as Weebly.com, Wix.com, Wordpress.com, a general Internet hosting company like GoDaddy, or even a free service like Google Sites. In those cases, they manage 95% of the technical part so that you don't have to, and in many cases it is automated through a system called a Content Management System. (If you weren't aware, what you used to have to do, and what some sites still do, is create and maintain a web site manually. That means you create the files on your computer and upload them to a server, and the server makes them available on the Internet.) What a Content Management System (CMS) does is make it a lot easier for someone without technical skills to build a web site, basically.

No matter how the web site is developed, sometimes there's still a need for a doctor.

That's where Webmaster Tools/Search Console comes in. There's even a Search Health function. Just like getting a check up.

Even though Webmaster Tools is now becoming Search Console, it's still good to know its history and context, and when you are searching for a job, an SEO job description likely will still list Webmaster Tools. You should put both terms on your resume and on your LinkedIn profile, once you gain some experience with these areas.

Before we dive in, I recommend looking at this Google Webmaster Tools Video: `https://www.youtube.com/watch?v=COcl6ax38IY` or `http://tinyurl.com/wmtoolsvid`.

If you like, take a look at the "revised" version, which is now called the Google Search Console: `https://www.youtube.com/watch?v=SoxU5kz15Kc` or `http://tinyurl.com/srchvid`.

Pretty minor differences, but it's worth watching and re-watching until it sinks in.

Connecting Webmaster Tools/Search Console

An entire book could be written about Webmaster Tools/Search Console, and there are articles and videos as well (don't forget to search Google for videos as you are learning SEO). The purpose is to get acquainted, to show you how it is just another Google account. As with Google Analytics, there may be some hoops to jump through to get connected. But getting more familiar with it is a good investment of your time, and you may want to go ahead and jump in.

To check it out, go to `https://www.google.com/webmasters/tools`. (If they update the link at some point, Google "Google Search Console" to find it again.) If you are signed in already with a Gmail address/Google account, it will go right to the site. If not, you may need to sign in with a Gmail account.

In general, what you end up with is the ability to add and manage properties. Think of it being like a real estate mogul. You add properties and keep track of them to make sure they are okay. You have someone working for you who lets you know if there's an issue.

This is what I see when I sign in:

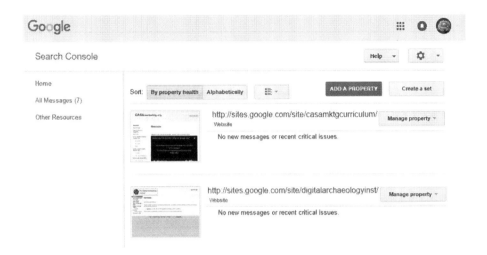

In general, you add a property, manage it, and sometimes check things. In some cases, there might be messages for you about something to look into; in other cases, they might e-mail you a message.

Let's try adding a property. First click the Add a Property button.

Remember, with the mindset of an SEO professional and digital marketing learner, always click on Learn More links and the little question marks in Google products. You'll be glad you did.

You could add a "conventional" web site, such as www.mytshirtshop.com, but for this exercise, we're adding a Google web site, which is a bit easier to connect. Type in your blog address as follows:

Add a property

Select the type of property you would like to manage. Learn more.

| Website ▾ | http://toddsmarketingblog.blogspot.com | ⑦ |

Add   Cancel

Click the Add button. In theory, Google will:

- Tell you that are already a verified owner. (If you aren't, make sure you're signed in to Webmaster Tools with the same Gmail you used to sign in and create your blog.)

- Allow you to verify with more than one format. It's not required, but in theory makes things more secure.

For web sites that aren't built on Google platforms (i.e., Google Sites and Blogger), you have to go through some kind of verification. Probably the easiest way to verify is to have Google analytics already set up on your site (Weebly, Wix, any other platform for that matter). Another verification process that might work is the domain provider option.

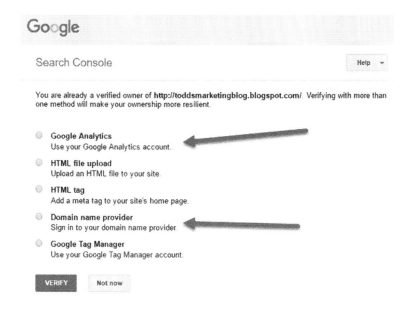

Until you want some extra credit from Dr. Kelsey and want to try one or more of those, click Not Now:

There will be many different options in Webmaster Tools, and I suggest exploring them. Messages is an interesting option. In theory you might get e-mails, but it's also a good idea to visit Webmaster Tools regularly (for example, you can set a recurring weekly reminder in Google calendar with an e-mail).

On the left side menu, there are things to explore like this:

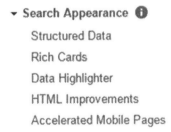

Oof! Don't feel bad if start to feel woozy when you see returns like this. Seriously, don't feel bad. It's doable. Want to know a little secret? Aside from the fact that I started out in rock 'n' roll, there's a secret to how I became an expert.

In theory, I'm an *industry* expert. I've been on TV, written a bunch of books, worked at small, medium, and large Fortune 500 and Fortune 50 companies— blah blah blah. But what I *haven't* done is tried to learn everything about every tool or piece of software I come across.

Remember that you don't have to learn everything. This entire digression has mostly been for the people who are intimidated by the last screenshot, with all the technical stuff. Suffice it to say, you can focus on the 10% of a program that is most learnable/doable for you now, and then learn the others later. Like successive coats of paint. That's what I do. That's the secret.

Okay, now back to Search Console!

Remember that as you explore it, you can click on Help in the interface.

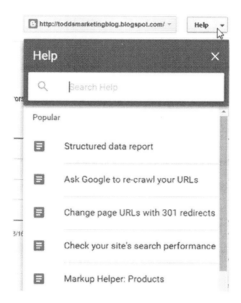

You can read through Google's own guide to it, which I recommend.

Google Search Console Help Center is at https://support.google.com/ webmasters?hl=en&authuser=0#topic=3309469 or http://tinyurl.com/ searchconsolehelp.

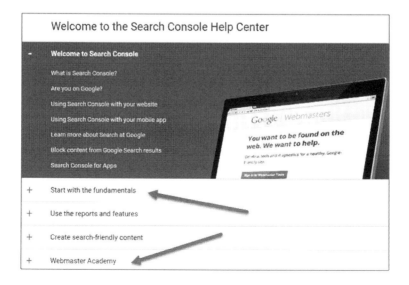

Well, look at all that interesting stuff! Start with the fundamentals. Webmaster Academy! That sounds good. Go for it!

My general recommendation is simply to explore—click on everything, read the help material, search for videos on Google, and get comfy with the SEO concept.

In some cases, as with Google Analytics, it takes time for data to come in before there is useful information or insights. But get it going and see what happens. (As with Google Analytics, if you want to see a live account, e-mail me at tekelsey@gmail.com and I will add you to a live Google Search Console so you can see some live data.)

It is worth noting that among other things, you can add other users. If you work for a company that already has a web site and a Google Webmasters account, there's likely a person who already has access. It's also possible they might not know they can add users. You can ask them nicely to add you to the account.

That kind of thing can also come up with Google Analytics as well. If you start working at a company that already has an account or you take on a new client, you might need to ask someone to invite you to an existing Google Analytics account.

News Flash: Checking Rank

Things change in SEO, as well as in other areas of digital marketing. You have to keep tuned in to the technology news in order to keep up. Sometimes you just end up discovering changes in the tools.

For example, at the time of this writing, Google introduced the ability to check your rank for individual keywords right in the Webmaster Tools/ Search Console. This is huge, because when you start doing search engine optimization, and you want to report on the results of your efforts to a client or manager, being able to report an increase in rank is helpful. But what if your keyword is not in the top 100? You're stuck if you're using some of the other tools.

Once you've connected all the pieces to your site, including Webmaster Tools/ Search Console, you can see which keywords people are typing into Google that lead them to your site.

To do this, go into Search Analytics:

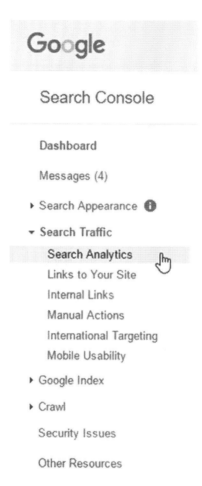

You will see something like this:

Search Analytics

Analyze your performance on Google Search. Filter and compare your results to better understand your user's search patterns. Learn more.

If you click the Position checkbox:

You should see the position of the search queries, meaning their rank on Google.

Queries	Clicks ▼	Position	
zing enterprises ⌕	1	1.0	»
aluminum cans only sign ⌕	1	9.0	»
red zing ⌕	0	10.0	»
door swings open ⌕	0	45.3	»
traffic signs in the workplace ⌕	0	80.0	»

This example tracks the Zing Green Safety Products company. The site has been around long enough that many people type the name of the company into Google. The "Zing Enterprises" phrase is number 1. This is the Holy Grail; the top rank in Google. It's a little easier for your brand name. Then further down we see a more conventional result—the term "traffic signs in the workplace" has a rank of 80. That means you have to scroll through 80 search results before you see Zing come up *organically*. You might see an ad on the very first page (created via AdWords), but the organic search result is number 80. An SEO-minded person would try to improve the rank for that keyword. With what? Content.

It takes time, but that's the idea.

Even though it's great that Google now allows you to check rank/position in the Search Console, don't be disappointed if you start a new web site and it doesn't show *any* keywords. It takes time, but it's worth the effort. The more traffic you can get to your web site using organic search results (i.e., whereby you are not paying for it), the better. Most companies use a combination of SEM and SEO to get traffic to their sites.

A realistic way to report on the success of content development efforts related to search engine optimization is to learn about Google Analytics, including how to measure traffic in general. Ad campaigns and content can both increase traffic. They can increase the amount of time people spend on the site as well. As you build your web site, look more closely at search engine optimization and engage in *active* SEO. When it comes time to show results, working toward increasing your rank is one way to get there. Another way is to study Google Analytics, which can tell you which keywords are driving traffic and how much of your traffic is coming from organic search (in the form of referrals).

I've jumped ahead to some advanced stuff in this section, but when you start doing SEO, the Search Console is a good tool to have. It is especially helpful with a new web site or smaller business that might not place on the top 100, and where you can't measure the results from your efforts on one of the larger, paid search rank tools (until they change it). Google Search Console may be your best friend in the hot seat.

Take time to learn Google Analytics as well. If it isn't crystal clear, it's important to be aware of how the various core areas of digital marketing relate. In the last several paragraphs, you've seen how SEO and SEM relate to Google Analytics, Search Console, etc.

If you haven't yet, check out my *Introduction to Google Analytics, Introduction to Search Engine Marketing and AdWords* and *Introduction to Social Media Marketing* books.

Conclusion

Congratulations on getting your feet wet with indexing and Webmaster Tools/Search Console. Remember, don't worry about absorbing everything, just learn how to explore. You can and will learn what you need to know over time.

Best wishes!

Keeping Up with Changes

This is perhaps the shortest chapter of any book I've written, and the main point is to emphasize that one of the parts of SEO is to *keep up with changes*.

It also mean looking critically at resources that might be out of date, including this book! Depending on when you read it of course. To balance that approach, I've tried to make this book more about learning how to learn than attempting to explain "eternal" SEO best practices. There are no practices that are eternal. Except maybe one—quality content!

Outside of looking for quality content, Google evolves and changes its algorithms. The entire SEO industry responds with workshops, blogs, webinars, seminars, conventions, and panel discussions. Sometimes Google makes announcements and sometimes things change quietly and people who are watching closely notice them. It's just the way things go.

Part of the reason that Google doesn't always disclose exactly how things work, is because in the past, present, and probably in the future, there will always be people who try to capitalize on exactly how things work to manipulate the system. They take it to an extreme and get good results for a while, but Google eventually catches up with them. Google's algorithmic changes have also caused legitimate businesses with the proper approach to have to revamp everything. Even if you're pursuing a particular business model that's content related and relies on search engine rankings, an algorithmic change can have a large impact on your results.

© Todd Kelsey 2017
T. Kelsey, *Introduction to Search Engine Optimization*, DOI 10.1007/978-1-4842-2851-7_8

For example, this is a report from a mid-2015 news article.

Google's "phantom" algorithm update hits web sites:

- Search change costs sites real money

- Web sites were not expecting the change

- One site loses 22 percent of traffic overnight

Check out `http://www.cnbc.com/2015/05/13/ntom-algorithm-update-hits-websites.html` or `http://tinyurl.com/goog-phantom`.

If you still don't believe me, look at Moz's article entitled, "Google Algorithm Change History" at `https://moz.com/google-algorithm-change`.

Sometimes Google acknowledges these issues and sometimes they don't.

The Quality Update — May 3, 2015
After many reports of large-scale ranking changes, originally dubbed "Phantom 2", Google acknowledged a core algorithm change impacting "quality signals". This update seems to have had a broad impact, but Google didn't reveal any specifics about the nature of the signals involved.

The Quality Update: Google Confirms Changing How Quality Is Assessed, Resulting In Rankings Shake-Up (SEL)

Google's 'phantom' algorithm update hits websites (CNBC)

Perhaps the biggest one in recent memory—which is still worth being aware of because you might be the one to explain its importance to a business or organization—is called "mobilegeddon". This was when Google (with ample public notice) started penalizing sites that were not mobile friendly.

Mobile Update AKA "Mobilegeddon" — April 22, 2015
In a rare move, Google pre-announced an algorithm update, telling us that mobile rankings would differ for mobile-friendly sites starting on April 21st. The impact of this update was, in the short-term, much smaller than expected, and our data showed that algorithm flux peaked on April 22nd.

Finding more mobile-friendly search results (Google)

7 Days After Mobilegeddon: How Far Did the Sky Fall? (Moz)

Suffice it to say that it's a good idea to keep up with changes. How do you do that? Read on.

Keep Up with Industry Resources and Blogs

Sometimes all it takes is reading a blog post. Sometimes it's helpful to read a book or attend a webinar. The best thing to do is get familiar with some of the industry resources out there. Here are a few that have been mentioned. Some are tools that also include news and others are news sites that have tools as well as training materials (and certifications, as you'll learn in the next chapter).

- www.searchenginenews.com

- www.moz.com

- www.semrush.com

Of course, try searching for "SEO news" to see what comes up in Google.

In general, the companies and sites that are focused on SEO news, such as the ones mentioned previously, are fairly current. They may be the ones you go to or hear from about the latest changes. There are also a host of blog posts from a variety of sources that will come up in a search, but do not reflect the latest changes in Google. For example, you might find a book, blog post, or article that talks about meta keywords. These used to be more important, but now they are less so. You need to look for information with a critical eye.

If you want my personal advice, go right to moz.com and subscribe to their blog at https://moz.com/blog.

Never miss a beat. Get Moz Blog
email updates daily in your inbox

Subscribe to Email Updates

You'll learn something, you'll keep updated, and you'll be glad you did.

I also recommend exploring *Search Engine News*. At the time of this writing, they have a $1 trial. I learned about them through an agency I worked for. The owner found their information helpful enough that it was worth paying for. See www.searchenginenews.com.

Conclusion

That's it. Really. Keeping up with SEO trends is really important. Learning the best there is to know about SEO, but not keeping up with changes is a recipe for disaster. Get plugged in and be aware of changes and get expert opinions on the relative importance of changes. That's valuable and it will help you do your job better.

Exploring SEO Certification

As you're learning SEO—whether you are adding this skill to an existing job, or you're learning this skill to boost your chances of getting a job—certification is something to seriously consider. In teaching the core areas of digital marketing, I've tried to integrate certification into as many classes and books as possible, and SEO is no different, although the "certification landscape" is not as clear as with search engine marketing, where you have a clear, recognized certification with AdWords, and with Web Analytics, where Google Analytics is free, accepted, and well-known.

For fun, I suggest that you take the following keywords and enter them into a job search in LinkedIn to see what you find:

- Search engine optimization
- SEO
- Google Analytics
- Google AdWords
- Google Analytics Certification
- HubSpot

© Todd Kelsey 2017
T. Kelsey, *Introduction to Search Engine Optimization*, DOI 10.1007/978-1-4842-2851-7_9

For additional inspiration, you might also want to go on salary.com or indeed.com and look at some of their salary tools. Don't forget LinkedIn. Check the site mid-January of each year for their post on the top 25 skills. Here is their post as of January of 2016, reflecting 2015 data.

https://blog.linkedin.com/2016/01/12/the-25-skills-that-can-get-you-hired-in-2016

or

http://tinyurl.com/2015-top25

There is SEO, at number 4. You made the right choice in reading this book! Look how it is grouped with SEM. Makes sense, right? What's also interesting is that marketing campaign management is number 3. That means of any skill to get people hired. In the world. That means there's increasing need to gain experience and be able to manage campaigns, like the flow through content, social, AdWords, and analytics. Does that sound familiar?

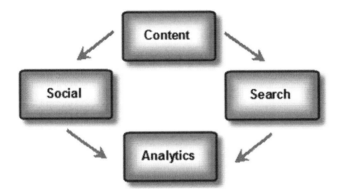

Hopefully some of the discussion is helping you understand that SEO, like other core digital marketing areas, is connected to other areas. For example, we looked at how Google *Analytics* can be used to measure the impact of *content* used for SEO. You also learned how *content* can be posted to *social media*. And so on.

The reason I developed the CASA model was not to promote myself, but to try and find a simple way to navigate digital marketing, and to introduce some of the relationships. It's not "my" tool, it's "our" tool, and it's a suggested mindset.

In light of some of these points, the next sections introduce two options for certification that I've tried, tested in classrooms and *strongly* recommend—HubSpot Certification and Search Engine News.

HubSpot Certification

HubSpot (www.hubspot.com) is cool. The quality of the learning material and the tools is top-notch. They practice what they preach, and it's very likely that you'll run across their blog posts from time to time. They are trying to promote themselves with the concept of "inbound marketing". They do it in a way that reflects their core philosophy and best practices that they encourage—make your content relevant and helpful, and when you do, customers will follow.

When you visit their site, you'll see something like this.

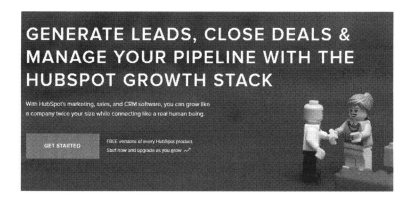

You may want to just go ahead and try some of their products, just to get familiar. They have a free CRM (customer relationship management) tool that can be used to track customers and relationships. Ultimately, they have tools that can help with inbound marketing, which is the process of doing everything you can to get people to visit your web site, using good *content*—and then measuring and optimizing the process.

If SEO is the process of optimizing a web site for higher rankings on Google, and if good quality content is the best way to do it, then inbound marketing is taking things to the next level. It's the way you can do "content marketing" and work on not just getting higher search engine rankings, but look at ways to help generate and track sales using a platform like HubSpot. Sometimes they call it "marketing automation". (If you want to learn more about marketing automation, HubSpot has a lot of material, and if you want a conversational introduction to HubSpot and its alternatives, see *Intro to Marketing Automation* by Todd Kelsey.)

In general, marketing automation, inbound marketing, is worth looking at, regardless of your business size, but it's probably better to look at it after you have a solid foundation of other things. Being aware of the possibilities is a good thing. HubSpot and its competitors are best for businesses that have a B2B focus, where you have leads, a sales force, and the need to help increase the conversion of leads into customers. Marketing automation helps a lot of this, based on the idea that 98 percent of the people who visit your site aren't ready to buy. If you've invested in getting them to visit (with SEO, SEM), how can you increase the conversion rate? The answer is lead nurturing, and marketing automation helps with that.

The reason I think the HubSpot learning material is a good foundation for SEO is because it places SEO in context, and the emphasis is on content. Not just doing it, but planning things so that it is relevant and helpful to people. Their training material and certification is excellent at giving a foundation in these concepts, and I strongly recommend it. It introduces things like Personas. If you want to explore the material and take a look at Personas, for example, go to `http://certification.hubspot.com/inbound-certifi-cation`. You can also just visit `certification.hubspot.com` and choose the inbound certification.

Go ahead and click Start the Course. Then follow the registration process.

Sign Up for HubSpot's Certification Programs

HubSpot

Language **English** Español Português Deutsch

First name

Last name

Email address

Phone

Company name

Website URL

What is your role?   ▼

How it works

1. **Sign up**

   Sign up for a complimentary HubSpot account. Get access to your certification library, study guides, and more helpful HubSpot tools. Already a paying customer? Just log in and look under the Academy tab.

2. **Take your classes**

   Each certification has its own preparatory classes and resources. Watch videos at your own pace and use your study guide to prepare you for your certification exam.

3. **Pass the exam**

   Register for and take your certification exam. Track your grade from within your HubSpot account and when you've passed, you'll join a global community of inbound marketing professionals.

As a side note, one of the things to file away at the border of SEO and overall content marketing is the idea of getting people's contact information in return for something. That's a fundamental principle in inbound marketing, and that's what's happening with the HubSpot registration process. They offer free certification, they ask for your contact information, and they try to send helpful resources along the way. They recognize that at some point you might be a potential customer or in the position to recommend HubSpot to someone else.

As part of the registration process, you pick a password. I recommend writing it down so you can more easily come back in later.

Password

Confirm password

☑ I agree to the HubSpot Terms of Service

Submit

You'll end up seeing something like this, and I suggest choosing the Inbound material.

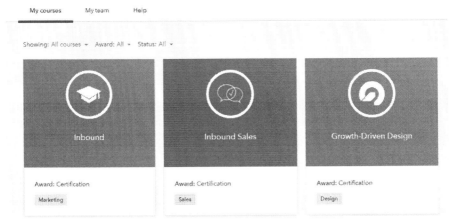

Then at this point what I suggest you do is bookmark the link to this page, the course hub, so you can come back to it later:

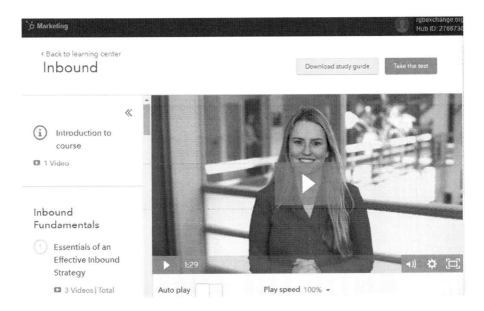

Download the study guide and then check out one or two of the videos just to see what it's like, maybe starting at the beginning. When you're ready, spend some time each day or each week going through the material. It's excellent, and it will lead to a HubSpot Inbound Certification that you can and should list on your resume. For fun, try doing a job search on LinkedIn for the term "hubspot".

I don't own any stock in HubSpot and I'm not an employee. I've just found the material to be *that good*. They do well with visuals and simplify things to be sensible. There are also good sections in the training manual that cover important concepts of digital marketing, like stages of awareness.

For example, at one point in the material you learn about the "buyer's journey," which helps you to realize that when you have content and advertising, it's not "one size fits all" and people are at different stages of the buying process. Knowing this, and strategically developing content based on this knowledge, is very powerful, and produces results.

It goes beyond SEO. If you're in an SEO position and the company is not doing inbound marketing, you might want to start the conversation. If the company is doing inbound, you will have some knowledge about how to collaborate. If you're selling digital marketing services, a client relationship might start with SEO, but you may be able to upsell the client on exploring inbound marketing and marketing automation.

The learning material also contains helps context, where the "tactical" things you are learning are placed in the context of an overall marketing strategy. There's certainly strategy within SEO, but it doesn't hurt to know how SEO fits into the larger picture, and how various aspects of digital marketing can combine to help generate revenue for a company. Knowledge is power—the power to help a company or a client succeed.

Don't worry about necessarily "mastering" all these stages—I recommend becoming familiar with them. It's knowledge that can help you as you grow in your skills and career.

One of my favorite parts of the HubSpot material that they do well is with buyer personas. In some ways you can simplify this to "listening to your customer".

Buyer personas represent a concept called *segmentation*, and as you learn more about SEM and SEO, you'll see how important it is. When you get to know your customers, when you go beyond the "one size fits all" mentality, you realize that not every customer is the same. Oftentimes there are segments. For example in a T-shirt business, part of the customer base might include coaches of local sports teams. Another segment might be bands that want to sell T-shirts to fans. When you start to learn more about your customers and tailor resources and marketing messages to their needs, the targeting becomes very powerful. It makes marketing and content more relevant to particular audiences, and therefore more valuable to them and to you.

The HubSpot material has an entire section in the study guide and videos that deals with SEO, and it's good reinforcement of best practices.

So have at it! Work on the certification, and best wishes in the process!

SEO Certification

SEO certification is a little harder to find, and generally what I've seen is that the companies that offer it are used to charging higher prices for training, webinars, and workshops, so it's out of reach for most of my students (although it's worth looking into if you're a working professional).

Search Engine News is one example: `https://www.searchenginenews.com/`

Conclusion

This chapter covered two possibilities for SEO certification and there may be more as time goes on. Whatever you do, make sure that the company that offers certification has pertinent and current material (if you're in doubt, ask them the last time they updated it). HubSpot is a pretty good, fairly stable company that should be around for a while, and Search Engine News is smaller, but also good, so we'll see. There are also many courses you can take online, (at Moz, some free, some not), but I like the idea of an actual certification, as a goal to work toward and to get managers' and recruiters' attention.

In the end, getting experience using actual tools is as, or more, important than the certification. But getting a certification can be an excellent way to supplement experience.

On that note, depending on what your goals are, if you are learning things on the side and intending to bring the skill into the workplace, if you are doing it for yourself, or seeking a new career, I do think that internships are a good thing to go for. Another excellent way to get experience is to look around for a non-profit organization in your area and see if they want some free or discounted help working with SEO and/or content marketing. Oftentimes they prioritize their budget to direct mission, and may welcome having some help on their web site.

If you're just starting out on your career, another issue to consider is the question of whether to seek a job as a dedicated SEO or look for a job where you wear multiple hats. A dedicated role focuses solely on SEO, whereas a multiple hats role might include SEO, but also include other areas. My general advice is to seek a multiple hats role, which is a little easier to break into, including by through an internship. This can help you get experience in each of the different areas. Positions such as "digital marketing coordinator" and "Internet marketing coordinator" usually indicate the multiple hats role. You can search keywords like SEO or Google Analytics and often find them mentioned in the context of a multiple hats role.

Then, if you particularly like SEO, you might want to find a dedicated role. Depending on the size of the company, it might be "all technical tweaks," but there might also be some room to get involved with content, and you can ask those questions in an interview. A company or organization or potential client might know that "they want to do SEO," but not really be aware about things like inbound marketing. With your fancy new HubSpot certification, you might be in a position to help them grow. This is also true if you are already in a marketing department and people are talking about doing SEO. You might bring inbound marketing into the conversation or raise the possibility of things like creating a company-wide task force to help strategically identify and develop content that could be used to help boost SEO rank.

The world's your oyster and I wish you the best. Feel free to connect with me on LinkedIn if you like (http://linkedin.com/in/tekelsey).

Thanks for reading this book, and best wishes in your career!

Special Request   Thank you for reading this book. If you purchased this book online, please consider going on where you purchased it and leaving a review. Thanks!

I

Index

© Todd Kelsey 2017
T. Kelsey, *Introduction to Search Engine Optimization*, DOI 10.1007/978-1-4842-2851-7

Get the eBook for only $5!

Why limit yourself?

With most of our titles available in both PDF and ePUB format, you can access your content wherever and however you wish—on your PC, phone, tablet, or reader.

Since you've purchased this print book, we are happy to offer you the eBook for just $5.

To learn more, go to http://www.apress.com/companion or contact support@apress.com.

Apress®

Made in the USA
Lexington, KY
19 January 2018